Genetics in Liver Diseases

FALK SYMPOSIUM 156

Genetics in Liver Diseases

Edited by

H.E. Blum
University Clinic Freiburg
Freiburg
Germany

D.W. Cox
University of Alberta
Edmonton
Canada

D. Häussinger
University Clinic Düsseldorf
Düsseldorf
Germany

P.L.M. Jansen
Academic Medical Center
Amsterdam
The Netherlands

G.A. Kullak-Ublick
University Hospital Zürich
Zürich
Switzerland

Proceedings of the Falk Symposium 156 held in Freiburg, Germany, October 8–9, 2006

Springer

Library of Congress Cataloging-in-Publication Data is available.

ISBN-13 978-1-4020-6392-3

Published by Springer,
PO Box 17, 3300 AA Dordrecht, The Netherlands

Sold and distributed in North, Central and South America
by Springer,
101 Philip Drive, Norwell, MA 02061 USA

In all other countries, sold and distributed
by Springer,
PO Box 322, 3300 AH Dordrecht, The Netherlands

Printed on acid-free paper

Printed and bound in Great Britain by MPG Books Limited, Bodmin, Cornwall.

Contents

List of principal contributors

HE Blum
Department of Medicine II
University Hospital
Hugstetter Strasse 55
D-79106 Freiburg
Germany

MA Buendia
Oncogenesis and Molecular Virology Unit
INSERM U579
Institut Pasteur
28, rue du Dr. Roux
F-75724 Paris Cedex 15
France

PT Clayton
Biochemistry Research Group
Clinical and Molecular Genetics Unit
UCL Institute of Child Health
30 Guilford Street
London, WC1N 1EH
UK

C Colombo
Department of Pediatrics, CF Center
Fondazione IRCCS 'Ospedale Maggiore Policlinico, Mangiagalli e Regina Elena'
University of Milan
I-20122 Milan
Italy

DW Cox
Department of Medical Genetics
8-39 Medical Sciences Building
University of Alberta
Edmonton
Alberta T6G 2H7
Canada

P Ginès
Liver Unit, Hospital Clinic
University of Barcelona
Villaroel 170
E-08036 Barcelona
Spain

S Gupta
Albert Einstein College of Medicine
Ullmann Building, Room 625
1300 Morris Park Avenue
Bronx, NY 10461
USA

PLM Jansen
Liver Center
Department of Hepatology and Gastroenterology, AMC
Meibergdreef 9
NL-1105 AZ Amsterdam
The Netherlands

GA Kullak-Ublick
Division of Clinical Pharmacology and Toxicology
Department of Internal Medicine
University Hospital
CH-8091 Zurich
Switzerland

F Lammert
University Hospital Bonn
Department of Internal Medicine I
Sigmund-Freud-Str. 25
D-53105 Bonn
Germany

UA Meyer
Biozentrum, University of Basel
Klingelbergstrasse 50–70
CH-4056 Basel
Switzerland

A Pietrangelo
Center for Hemochromatosis and Hereditary Liver Diseases
Department of Internal Medicine
University of Modena and Reggio Emilia Policlinico
Via del Pozzo 7
I-41100 Modena
Italy

J Prieto
Clinica Universitaria de Navarra
Departamento de Medicina Interna
Apartado 192
E-31080 Pamplona
Spain

EA Roberts
Division of Gastroenterology, Hepatology and Nutrition
Room 8257, Black Wing
The Hospital for Sick Children
555 University Avenue
Toronto
Ontario, M5G 1X8
Canada

T Sveger
Department of Pediatrics
Lund University
University Hospital
SE 205 05 Malmö
Sweden

R Thimme
Department of Medicine II
University Hospital Freiburg
Hugstetter Strasse 55
D-79106 Freiburg
Germany

S Thorgeirsson
Laboratory of Experimental Carcinogenesis
Center for Cancer Research
National Cancer Institute, NIH
37 Convent Drive, MSC 4262
Building 37, Room 4146A
Bethesda, MD 20892-5688
USA

List of chairpersons

HE Blum
Department of Medicine II
University Hospital
Hugstetter Strasse 55
D-79106 Freiburg
Germany

DW Cox
Department of Medical Genetics
8-39 Medical Sciences Building
University of Alberta
Edmonton
Alberta T6G 2H7
Canada

D Häussinger
Department of Gastroenterology and Hepatology
University Clinic Düsseldorf
Moorenstr. 5
D-40225 Düsseldorf
Germany

PLM Jansen
Liver Center
Department of Hepatology and Gastroenterology, AMC
Meibergdreef 9
NL-1105 AZ Amsterdam
The Netherlands

GA Kullak-Ublick
Division of Clinical Pharmacology and Toxicology
Department of Internal Medicine
University Hospital
CH-8091 Zurich
Switzerland

W Stremmel
Department of Internal Medicine IV
University Clinic
Im Neuenheimer Feld 410
D-69120 Heidelberg
Germany

Preface

The Falk Symposium No. 156 'Genetics in Liver Diseases' was Part II of the XIII International Falk Liver Week, held in Freiburg, 6–11 October, 2006 with Prof. Gustav Paumgartner, Munich as president, which also included Falk Symposium No. 155 'Bile Acids: Biological Actions and Clinical Relevance' and Falk Symposium No. 157 'Chronic Hepatitis: Metabolic, Cholestatic Viral and Autoimmune'.

The symposium 'Genetics in Liver Diseases' was organized in three sessions. Session I was devoted to basic aspects of genetics of liver diseases and was introduced by a state-of-the-art lecture on 'Pharmacogenetics' by Prof. Kullak-Ublick, Zürich, and included overviews of microarray analyses of liver diseases, genetics of gallstone formation and of viral hepatitis B and C as well as the molecular pathogenesis of hepatocellular carcinoma (HCC). Special highlights of Session I were the award of the 'International Hans Popper Prize' to Prof. Gines, Barcelona, and his 'Hans Popper Award Lecture' on 'Circulatory and renal failure in cirrhosis'.

Session II was introduced by a state-of-the-art lecture on 'Hereditary hemochromatosis: the genes and the disease' by Prof. Pietrangelo, Modena, and focussed on our current knowledge of the molecular pathogenesis, diagnosis, treatment and prevention of Wilson disease, α_1-antitrypsin deficiency, porphyrias and cystic fibrosis. The final Session III, introduced by a state-of-the-art lecture on 'Genetics of cholestasis' by Prof. Jansen, Amsterdam, addressed new developments in the field of screening strategies for hereditary liver diseases, experimental and clinical issues of gene therapy, especially for HCC, as well as stem cell therapy.

In the tradition of the previous International Falk Liver Weeks in Basel and Freiburg, Falk Symposium No. 156 again provided an exciting overview of the state of the art of the genetics and molecular pathogenesis of hereditary liver diseases as well as the impact on their diagnosis, treatment and prevention, including gene and stem cell therapy, presented by an international faculty of outstanding scientists and clinicians. Their contributions should be of interest to all who attended the symposium but also to all those interested in recent developments in the field of genetics as well as molecular aspects of liver diseases, including their current and future relevance for patient care.

H. E. Blum

Section I
Basic aspects

Chair: HE BLUM and D HÄUSSINGER

1
State-of-the-Art Lecture I: Pharmacogenetics

N. CORTI and G. A. KULLAK-UBLICK

INTRODUCTION

The individual response to drug therapy may vary greatly, making the success of a given treatment strategy difficult to predict, particularly because multiple factors influence drug response. In addition to age, sex, comorbidity, renal and hepatic function, the role of genetic factors that determine an individual's response to drug treatment is increasingly being realized. Genetic differences influencing the pharmacokinetics of a drug may exist at the level of drug metabolism determined by several well-characterized phase I and phase II enzymes, with cytochrome P450 being the most important metabolizing system. Drug disposition is influenced not only by metabolizing enzymes but also by transport proteins, that play an important role in the processes of drug absorption, distribution and excretion. In addition to pharmacokinetic determinants pharmacodynamic effects of a drug may be determined by genetic differences in the drug target. Although to date an increasing number of genetic polymorphisms (SNP) in single genes has been associated with variability of protein function and drug disposition, dose–response is generally influenced by a whole spectrum of genes, making analysis of a single gene insufficient to explain the variability of a pharmacological effect. Despite this complex interaction of factors determining drug response, several examples of direct genotype–phenotype relations have been discovered, that have led to the development of routine diagnostic tools for the individualization of drug dosing regimens.

DRUG-METABOLIZING ENZYMES

Drug-metabolizing enzymes are subdivided into two categories: phase I and phase II enzymes. Phase I enzymes are responsible for the oxidative degradation (oxidation, reduction, hydrolysis) of endogenous substances (e.g. steroids) and xenobiotics. Phase II enzymes render the compounds more water-soluble mainly by glucuronidation, sulphation, acetylation and methylation.

The most important phase I enzyme system is cytochrome P450 (CYP), that metabolizes the majority of pharmacological compounds (50–60%). As many as 57 CYP genes have been characterized; among these only the enzymes encoded by CYP1, CYP2 and CYP3 genes seem to contribute to the metabolism of drugs[1]. Almost all of the CYP isoforms possess genetic polymorphisms, the CYP2 family exhibiting the highest genetic variability with a distinct clinical impact. CYP2D6 is the best-characterized isoenzyme and is responsible for the metabolism of most antipsychotics and antidepressants and several antiarrhythmics. In 6–10% of Caucasians and 1% of Asians the enzymatic activity is completely absent, resulting in a 'poor metabolizer' (PM) phenotype and leading to increased exposure and toxicity of drugs metabolized by CYP2D6[2]. Some individuals possess multiple copies of the gene expressing the ultra-rapid metabolizer (UM) phenotype resulting in subtherapeutic plasma concentrations at normal doses[3]. About 3% of Caucasians carry the CYP2C9 or CYP2C19 PM phenotype predisposing them to an increased anticoagulatory effect of s-warfarin and increased toxicity of the antiepileptic drug phenytoin. For many drugs conjugation by phase II enzymes represents the main metabolic pathway, although these are not as well characterized as the oxidative enzymes. Nevertheless a few important polymorphisms have been identified. Thiopurine *S*-methyltransferase (TPMT) catalyses the *S*-methylation of azathioprine (AZA), 6-mercapto-purine (6-MP) and thioguanine. TPMT activity exhibits genetic polymorphisms in 10% of Caucasians, with 1/300 individuals having a complete deficiency that leads to an increased risk for fatal myelosuppression[4]. Toxicity of the antituberculostatic agent isoniazid has been associated with a reduced activity of the enzyme *N*-acetyltransferase-2. Life-threatening diarrhoea and leucopenia is observed in up to 25% of patients receiving the anticancer drug irinotecan. These adverse effects have been related to polymorphisms in the coding as well as in the promoter region of the uridine diphosphate–glucuronosyltransferase (UGT)-1A1, which is responsible for the glucuronidation and detoxification of irinotecan[5].

DRUG TRANSPORTERS

An increasing list of transport proteins have been found to play an important role in regulating drug uptake from intestine and sinusoidal blood into the liver, in regulating and limiting drug distribution to different body compartments such as the central nervous system, and in controlling drug efflux into bile and renal tubule[6]. Thus genetic polymorphisms in human membrane transporters may also contribute to interindividual differences in drug response. The transport proteins described to date that are involved in transmembrane drug transport include the ATP-binding cassette (ABC) transporters MDR1 (multidrug-resistance protein 1), BSEP (bile salt export pump), MRP1, 2, 3 and 4 (multidrug resistance associated protein 1, 2, 3 and 4) and BCRP/ABCG2 (breast cancer resistance protein) found in the basolateral or canalicular membrane of liver cells and some of them in the kidney. Organic anion and cation transporters such as the organic anion transporting polypeptides (OATP) are mainly responsible for drug uptake into the hepatocyte, and the organic anion and cation tranporters (OAT/OCT) for the renal tubular secretion and reabsorption of drugs. Multiple SNP and haplotypes have been characterized in the MDR1 gene (*ABCB1*) and its gene product p-glycoprotein (p-gp), which has been recognized to play a primary role in drug disposition[7]. Although data on the functional effect of the variant MDR1 alleles are not always conclusive, and are sometimes contradictory, the SNP 3435TT in exon 26 was associated with reduced intestinal expression of p-gp. Oral bioavailability of digoxin, the most extensively studied p-gp substrate, was significantly higher in individuals carrying the genotype 3435CC[8]. Enhanced p-gp activity was found in subjects with the 3435TT/G2677T genotype, causing a 40% reduction in AUC values following oral administration of the p-gp substrate fexofenadine, when compared to the 3435CC genotype[9]. MRP2 plays a central role in the biliary excretion of organic anions such as many anticancer drugs. Recently a mutation in exon 10 of the MRP2 gene was shown to confer loss of transport activity, leading to methotrexate overdose with renal failure[10]. The organic anion transporting polypeptide 1B1 (OATP1B1, *SLCO1B1*) and multidrug resistance-associated protein 2 (MRP2/*ABCC2*), are thought to be the major transporters determining the pharmacokinetics of pravastatin in humans. A genetic polymorphism in the OATP1B1 (*SLCO1B1*) gene led to a reduced total and renal clearance of pravastatin in healthy Japanese individuals[11].

It is now evident that hepatocellular transport proteins are also important in determining the exposure of the liver to drugs, and they may therefore be involved in the onset of drug-induced liver injury[12,13]. Reduced canalicular efflux transport protein activity can lead to intracellular bile acid and drug accumulation, resulting in liver cell damage. A polymorphism in BSEP (*ABCB11*) with a valine-to-alanine exchange in the highly conserved amino acid position 444 of the BSEP protein has been associated with a decreased hepatic content of BSEP[14] and with a threefold increased risk of developing cholestatic drug side-effects[15].

DRUG TARGETS

Genetic polymorphisms in drug targets can have a major influence on the effect of a drug. To date more than 20 examples of drug targets with genetic variability that directly influences drug response have been identified. The anticoagulatory effect of warfarin is determined on the one hand by the genotype and activity of the metabolizing enzyme CYP2C9 and on the other hand by the genetic variability of the target enzyme vitamin K epoxide reductase (VKOR)[16]. Homozygous mutations in the β_1-adrenergic receptor reduce sensitivity to beta-blocking agents in the treatment of arterial hypertension[17], and a correlation has also been seen between mutations in the β_2 adrenergic receptor and response to oral β_2-agonists[18]. Interindividual response and increased toxicity of psychotropic substances such as antidepressants and antipsychotics can be defined by the genotype of the corresponding receptor type[19,20]. Recently several examples of genetic variability in proteins targeted by monoclonal antibodies have been found to cause an insufficient drug effect. A study with 200 patients suffering from Crohn's disease showed a significant association between the biological response to infliximab, a chimeric antibody binding tumour necrosis factor-α, and a single nucleotide polymorphism in the FCGR3A gene, which codes for a receptor for the Fc fragment of IgG (FccR) on natural killer cells and macrophages[21]. In breast cancer therapy the monoclonal antibody trastuzumab inhibits the proliferation of tumour cells overexpressing HER2, a transmembrane tyrosine kinase receptor expressed on mammary carcinoma cells. The copy number or expression levels of the *HER2* gene is highly associated with the therapeutic efficacy of trastuzumab[22].

CONCLUSION

The discovery of further genotype–phenotype associations will define the role of pharmacogenetics in future clinical practice. Genotype-based prescribing could improve response rate and decrease the socioeconomic burden of adverse events. Nevertheless the effects of most drugs are determined by several proteins, and composite genetic polymorphisms in multiple genes coupled with non-genetic factors often have a high impact on individual drug response. New strategies are therefore needed to identify, for a given drug, the relevant genes; these include: (a) genome-wide haplotype mapping, (b) gene expression analyses, (c) proteomic methods, and (d) candidate gene approaches based on known pharmacokinetic and pharmacodynamic factors. These approaches, coupled with evolving biostatistical models, will elucidate polygenic determinants of drug response. Clinical validation of these polygenic models requires large clinical trials of uniformly treated and systematically characterized patients, high-throughput genomic methods, and sophisticated bioinformatic analyses. These studies could yield an exciting new panel of molecular diagnostics (i.e. genotypes) that can be employed to improve drug therapy by reducing toxicity and increasing efficacy.

References

1. Daly AK. Pharmacogenetics of the cytochromes P450. Curr Top Med Chem. 2004;4:1733–44.
2. Zanger UM, Raimundo S, Eichelbaum M. Cytochrome P450 2D6: overview and update on pharmacology, genetics, biochemistry. Naunyn Schmiedebergs Arch Pharmacol. 2004;369:23–37.
3. Johansson I, Lundqvist E, Bertilsson L, Dahl ML, Sjoqvist F, Ingelman-Sundberg M. Inherited amplification of an active gene in the cytochrome P450 CYP2D locus as a cause of ultrarapid metabolism of debrisoquine. Proc Natl Acad Sci USA. 1993;90:11825–9.
4. McLeod HL, Siva C. The thiopurine *S*-methyltransferase gene locus – implications for clinical pharmacogenomics. Pharmacogenomics. 2002;3:89–98.
5. Ando Y, Saka H, Ando M et al. Polymorphisms of UDP-glucuronosyltransferase gene and irinotecan toxicity: a pharmacogenetic analysis. Cancer Res. 2000;60:6921–6.
6. Mizuno N, Niwa T, Yotsumoto Y, Sugiyama Y. Impact of drug transporter studies on drug discovery and development. Pharmacol Rev. 2003;55:425–61.
7. Pauli-Magnus C, Kroetz DL. Functional implications of genetic polymorphisms in the multidrug resistance gene MDR1 (ABCB1). Pharm Res. 2004;21:904–13.

8. Hoffmeyer S, Burk O, von Richter O et al. Functional polymorphisms of the human multidrug-resistance gene: multiple sequence variations and correlation of one allele with P-glycoprotein expression and activity in vivo. Proc Natl Acad Sci USA. 2000;97:3473–8.
9. Kim RB, Leake BF, Choo EF et al. Identification of functionally variant MDR1 alleles among European Americans and African Americans. Clin Pharmacol Ther. 2001;70:189–99.
10. Hulot JS, Villard E, Maguy A et al. A mutation in the drug transporter gene ABCC2 associated with impaired methotrexate elimination. Pharmacogenet Genom. 2005;15:277–85.
11. Nishizato Y, Ieiri I, Suzuki H et al. Polymorphisms of OATP-C (*SLC21A6*) and OAT3 (*SLC22A8*) genes: consequences for pravastatin pharmacokinetics. Clin Pharmacol Ther. 2003;73:554–65.
12. Stieger B, Fattinger K, Madon J, Kullak-Ublick GA, Meier PJ. Drug- and estrogen-induced cholestasis through inhibition of the hepatocellular bile salt export pump (Bsep) of rat liver. Gastroenterology. 2000;118:422–30.
13. Funk C, Ponelle C, Scheuermann G, Pantze M. Cholestatic potential of troglitazone as a possible factor contributing to troglitazone-induced hepatotoxicity: *in vivo* and *in vitro* interaction at the canalicular bile salt export pump (Bsep) in the rat. Mol Pharmacol. 2001;59:627–35.
14. Meier Y, Pauli-Magnus C, Zanger UM et al. Interindividual variability of canalicular ATP-binding-cassette (ABC)-transporter expression in human liver. Hepatology. 2006;44:62–74.
15. Lang C, Meier Y, Stieger B et al. Mutations and polymorphisms in the bile salt export pump and the multidrug resistance protein 3 associated with drug-induced liver injury. Pharmacogenet Genom. 2007;17:47–60.
16. Oldenburg J et al. Congenital deficiency of vitamin K dependent coagulation factors in two families presents as a genetic defect of the vitamin K-epoxide-reductase-complex. Thromb Haemost. 2000;84:937–41.
17. Sofowora GG, Dishy V, Muszkat M et al. A common beta1-adrenergic receptor polymorphism (Arg389Gly) affects blood pressure response to beta-blockade. Clin Pharmacol Ther. 2003;73:366–71.
18. Lima JJ, Thomason DB, Mohamed MH, Eberle LV, Self TH, Johnson JA. Impact of genetic polymorphisms of the beta2-adrenergic receptor on albuterol bronchodilator pharmacodynamics. Clin Pharmacol Ther. 1999;65:519–25.
19. Hwu HG, Hong CJ, Lee YL, Lee PC, Lee SF. Dopamine D4 receptor gene polymorphisms and neuroleptic response in schizophrenia. Biol Psychiatry. 1998;44:483–7.
20. Kim DK, Lim SW, Lee S et al. Serotonin transporter gene polymorphism and antidepressant response. Neuroreport. 2000;11:215–19.
21. Louis E, El Ghoul Z, Vermeire S. Association between polymorphism in IgG Fc receptor IIIa coding gene and biological response to infliximab in Crohn's disease. Aliment Pharmacol Ther. 2004;19:511–19.
22. McKeage K, Perry CM. Trastuzumab: a review of its use in the treatment of metastatic breast cancer overexpressing HER2. Drugs. 2002;62:209–43.

2
Microarray analysis and liver diseases

S. S. THORGEIRSSON

INTRODUCTION

The notion that a disease condition may represent disruption of the normal network structures of an organ system through either genetic perturbations and/or deleterious environmental agents is a central tenet of the systematic approaches currently used to understand complex diseases[1]. Cancer is a complex disease caused by multiple spontaneous and/or inherited mutations that provoke extensive changes in gene and protein expressions that function in networks controlling critical cellular events[2]. A major challenge in cancer research is to characterize the complex molecular networks that drive hepatocellular carcinoma (HCC) development in order to provide both deeper understanding of cancer pathogenesis and more effective treatment modalities. This challenge has been facilitated by the emergence of novel high-throughput technologies, including DNA microarrays that can simultaneously detect the expression levels of thousands of genes. This new technology has been used successfully to predict clinical outcome and survival, as well as to classify different types of cancer[3].

Numerous studies have recently described gene expression profiles in human liver diseases, particularly HCC[4]. Most of these studies have concentrated on identifying genes associated with clinicopathological features (i.e. descriptive studies). However, some studies have correlated gene expression patterns with clinical outcome of liver patients (e.g. HCC survival or recurrence). It is therefore expected that microarray-based expression technologies may complement the pathology evaluation of liver biopsies[5]. Furthermore, appropriate application of gene expression analysis will facilitate the identification of therapeutic targets for patients with liver disease. However, several obstacles still need to be overcome before routine clinical application of diagnostic microarray-based methods becomes a reality. These include variations in gene expression in normal and diseased livers, as well as technical factors relating to experimental design and the huge amount of data that is generated[6].

Here we illustrate examples of how gene expression profiling of HCC can provide a better understanding of the molecular pathogenesis of the disease and how the application of an integrative functional comparative genomics approach can enrich the human gene expression database on HCC.

RESULTS AND DISCUSSION

Predicting survival

Gene expression patterns of 91 human HCC were used to define the molecular characteristics of the tumours and to test the prognostic value of the expression profiles[7]. Application of unsupervised classification methods led to the discovery of two distinctive subclasses of HCC highly associated with survival of patients. Cox proportional hazards survival analysis was also used to identify genes most strongly associated with survival. This method identified a limited number of genes that correctly predicted the length of survival, and provides a new molecular insight into the pathogenesis of HCC. The low survival subclass tumours were characterized by strong cell proliferation and anti-apoptosis gene expression signatures. Furthermore, the low survival subclass tumours had higher expression of genes involved in ubiquitination and histone modification, suggesting an aetiological involvement of these processes in accelerating the progression of HCC. The biological differences identified in the HCC subclasses are expected to provide an attractive source for the development of therapeutic targets for selective treatment of HCC patients.

Cancer stem cells in HCC

The variability in the prognosis of individuals with HCC suggests that this cancer may be composed of several distinct biological phenotypes. These phenotypes may be characterized by activation of different oncogenic pathways and/or from a different cell of origin. We have used a transcriptomic approach to provide insight into the cellular origin of HCC[8]. This approach involved the integration of gene expression data from rat fetal hepatoblasts and adult hepatocytes with HCC from human and mouse models. We discovered that individuals with HCC who shared a gene expression pattern with fetal hepatoblasts had a particularly poor prognosis. Included in the gene expression programme that distinguished this subtype from other types of HCC were markers of human hepatic oval cells, suggesting that HCC of this subtype may arise from hepatic progenitor cells. Further analyses of the 'hepatic progenitor gene network' revealed activation of AP-1 transcription factors, possibly indicating a key role for these transcription factors in the development of this newly identified HCC subtype.

Integrated functional genomics of HCC

As indicated earlier, gene expression profiling studies of various cancers have discovered consistent gene expression patterns associated with a histological or clinical phenotype and discovering subtypes of cancers previously unidentified with conventional technologies[9–15]. The research focus has now shifted towards identifying genetic determinants that are components of specific regulatory pathways altered in cancers, in order to increase discoveries of novel therapeutic targets. However, for a number of reasons[16], it is difficult to select a few candidate genes for further studies from extensive gene lists generated from gene expression profiling studies. This limitation has been partly overcome by applying cross-comparison of multiple gene expression data sets from human HCC and animal models, uncovering evolutionarily conserved gene expression patterns during HCC progression[16,17]. The success of a new experimental and analytical approach, comparative functional genomics, suggests that more integration of independent data sets will enhance our ability to identify key regulatory elements during HCC development.

Furthermore, cross-species comparisons of gene expression data from human cancer and mouse models indicated that gene expression signatures reflecting similar phenotypes in two species are conserved as genomic architectures, including *cis*-regulatory elements[17–19]. If gene expression signatures are conserved in evolutionarily-related species it will be reasonable to speculate that the transcription factors governing these signatures would also be conserved. Research along these lines is currently ongoing in a number of laboratories.

References

1. Hood L, Heath JR, Phelps ME, Lin B. Systems biology and new technologies enable predictive and preventative medicine. Science. 2004;306:640–3.
2. Thorgeirsson SS, Grisham JW. Molecular pathogenesis of human hepatocellular carcinoma. Nat Genet. 2002;31:339–46.
3. Pusztai L, Ayers M, Stec J, Hortobagyi GN. Clinical application of cDNA microarrays in oncology. Oncologist. 2003;8:252–8.
4. Thorgeirsson SS, Lee JS, Grisham JW. Molecular prognostication of liver cancer: end of the beginning. J Hepatol. 2006;44:798–805.
5. Thorgeirsson SS, Lee JS, Grisham JW. Functional genomics of hepatocellular carcinoma. Hepatology. 2006;43:S145–50.
6. Lee JS, Thorgeirsson SS. Genetic profiling of human hepatocellular carcinoma. Semin Liver Dis. 2005;25:125–32.
7. Lee JS, Chu IS, Heo J et al. Classification and prediction of survival in hepatocellular carcinoma by gene expression profiling. Hepatology. 2004;40:667–76.
8. Lee JS, Heo J, Libbrecht L et al. A novel prognostic subtype of human hepatocellular carcinoma derived from hepatic progenitor cells. Nat Med. 2006;12:410–16.
9. Alizadeh AA, Eisen MB, Davis RE et al. Distinct types of diffuse large B-cell lymphoma identified by gene expression profiling. Nature. 2000;403:503–11.
10. van de Vijver MJ, He YD, van't Veer LJ et al. A gene-expression signature as a predictor of survival in breast cancer. N Engl J Med. 2002;347:1999–2009.
11. Bullinger L, Dohner K, Bair E et al. Use of gene-expression profiling to identify prognostic subclasses in adult acute myeloid leukemia. N Engl J Med. 2004;350:1605–16.
12. Dave SS, Wright G, Tan B et al. Prediction of survival in follicular lymphoma based on molecular features of tumor-infiltrating immune cells. N Engl J Med. 2004;351:2159–69.
13. Lapointe J, Li C, Higgins JP et al. Gene expression profiling identifies clinically relevant subtypes of prostate cancer. Proc Natl Acad Sci USA. 2004;101:811–16.
14. Valk PJ, Verhaak RG, Beijen MA et al. Prognostically useful gene-expression profiles in acute myeloid leukemia. N Engl J Med. 2004;350:1617–28.

15. Roepman P, Wessels LF, Kettelarij N et al. An expression profile for diagnosis of lymph node metastases from primary head and neck squamous cell carcinomas. Nat Genet. 2005; 37:182–6.
16. Lee JS, Thorgeirsson SS. Comparative and integrative functional genomics of HCC. Oncogene. 2006;25:3801–9.
17. Lee JS, Chu IS, Mikaelyan A et al. Application of comparative functional genomics to identify best-fit mouse models to study human cancer. Nat Genet. 2004;36:1306–11.
18. Ellwood-Yen K, Graeber TG, Wongvipat J et al. Myc-driven murine prostate cancer shares molecular features with human prostate tumors. Cancer Cell. 2003;4:223–38.
19. Sweet-Cordero A, Mukherjee S, Subramanian A et al. An oncogenic KRAS2 expression signature identified by cross-species gene-expression analysis. Nat Genet. 2005;37:48–55.

3
Genetics of gallstone disease

F. LAMMERT and H. WITTENBURG

Gallstone disease is one of the major digestive diseases: 10–20% of Europeans and Americans carry gallbladder stones[1,2]. Each year an estimated 700 000 cholecystectomies are performed in the USA. Cholelithiasis incurs annual medical expenses of $6.5 billion in the USA and is the second most expensive digestive disease, exceeded currently only by gastro-oesophageal reflux disease. Epidemiological studies in high-risk populations indicate that gallstone formation is caused by multiple environmental genetic factors and their interactions[3]. Gallstone prevalence is highest in some American populations, in particular Amerindians[4]. Significant risk factors for gallstones are age, female gender, parity and a high-caloric, high-carbohydrate and low-fibre diet. Gallstones appear to be associated with the metabolic syndrome and its major components, such as hyperinsulinism, dyslipidaemia, and abdominal adiposity. Enterohepatic infections, specifically by *Helicobacter* species, have been implicated in cholelithiasis and cholecystitis[5].

Recent twin and family studies strongly support the relevance of a genetic component to gallbladder disease (GBD). Analysing data from a large Swedish study of 43 141 twin pairs, we determined that genetic influences are major contributors to the variation in symptomatic GBD[6]: Genetic factors accounted for 25%, shared environmental factors for 13%, and unique environmental factors for 62% of the phenotypic variance among twins. Mono- and oligogenic subtypes of cholesterol cholelithiasis, such as biliary lipid transporter deficiencies, appear to be rare (Table 1, Figure 1). One form of cholelithiasis (*GBD1*) that is characterized by occurrence of intrahepatic sludge and cholesterol gallbladder stones before the age of 40, as well as recurrence of symptoms after cholecystectomy, is associated with mutations in the *ABCB4* gene encoding the hepatocanalicular phosphatidylcholine floppase[7]. Familial hypobetalipoproteinaemia due to mutations of the *APOB* gene might also be associated with formation of cholesterol gallstones, which appear to result from increased cholesterol secretion as a compensatory mechanism to eliminate cholesterol not incorporated into ApoB-containing very low-density lipoproteins (VLDL)[8].

The first genome-wide scan for gallstone susceptibility loci was performed in the San Antonio Family Diabetes/Gallbladder Study (SAFGS). Using

Table 1 Oligogenes associated with cholelithiasis in humans

Gene	*Presumed subtype*	*Genomic location*	*Putative lithogenic mechanism*	*Concordant Lith loci from murine studies*[16]
CFTR (ABCC7)[11]	Bilirubin	7q31	Induced bilirubin cycling; increased deconjugation of biliary bilirubin	*Lith10* (Chr 6, cM 0–10)[a]
UGT1A1[12]	Bilirubin	2q37	Decreased hepatic conjugation of bilirubin	None detected
ABCB4 (GBD1)[b 7]	Cholesterol	7q21	Reduced biliary phospholipid secretion	None detected
ABCB11[19]	Cholesterol	2q24	Reduced biliary bile salt secretion	*Lith1* (Chr 2, cM 32–52)
APOB[8]	Cholesterol	2p24	Increased biliary cholesterol secretion secondary to decreased VLDL synthesis	None detected
CCKAR[20]	Cholesterol	4p15	Impaired gallbladder and small intestinal motility	*Lith13* (Chr 5, cM 10–45)
CYP7A1[21]	Cholesterol	8q11-q12	Impaired cholesterol catabolism into bile salts	None detected

[a]Name of the susceptibility locus (Chr, chromosome; cM, centiMorgan of the 95% confidence interval).

[b]Gallbladder disease locus.

Table 2 *LITH* genes associated with common polygenic predisposition to gallstones in humans

Gene	*Type of study*	*Genomic location*	*Putative lithogenic mechanism*	*Concordant Lith loci from murine studies*[16]
GBD2[b 9]	Linkage	1p36	Unkown	*Lith8* (Chr 4, cM 48–84)[a]
GBD3[9]	Linkage	1p34	Unknown	*Lith8* (Chr 4, cM 48–84)
APOB[22]	Association	2p24	Increased biliary cholesterol secretion secondary to decreased VLDL synthesis	None detected
CETP[23]	Association	16q21	Increased HDL catabolism	None detected
CYP7A1[22]	Association	8q11-q12	Impaired cholesterol catabolism into bile salts	None detected
LRPAP1[24]	Association	4p16	Increased hepatic uptake of cholesterol from chylomicron remnants	*Lith13* (Chr 5, cM 10–45)

[a]Name of the susceptibility locus (Chr, chromosome; cM, centiMorgan of the 95% confidence interval).

[b]Gallbladder disease locus.

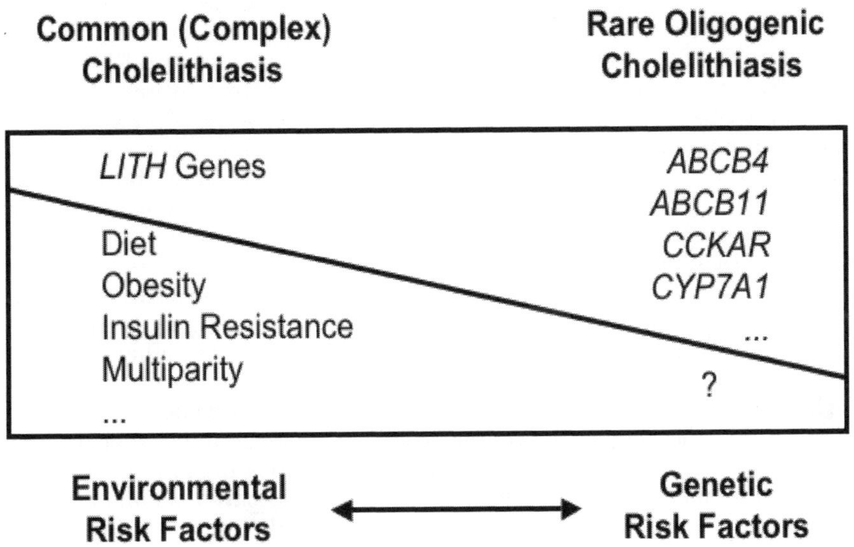

Figure 1 Schematic model of genetic and environmental factors contributing to 'oligogenic' and 'complex' cholelithiasis, respectively. As indicated by the diagonal, the proportion of genetic factors is likely to be smaller and the influence of environmental factors is probably larger in common 'complex' cholelithiasis compared to the rare forms of oligogenic gallstone formation. With permission from ref. 18

ultrasound-based information and a 10-cM gene map, Puppala et al.[9] employed multipoint variance components analysis to localize *GBD* susceptibility loci in 715 individuals from Mexican American families (Table 2). Two phenotypes were defined: (a) clinical or symptomatic GBD, the cases of cholecystectomies due to stones; and (b) total GBD, the clinical GBD cases plus the stone carriers newly diagnosed by ultrasound. Five metabolic risk factors were defined: increased waist circumference, hypertriglyceridaemia, low high-density lipoprotein (HDL) cholesterol, hypertension, and high fasting glucose. A metabolic syndrome risk score for a given individual was used as a single composite covariate in the genetic analyses. After accounting for the effects of age, gender, and the metabolic syndrome risk score, stronger linkage signals were detected for the symptomatic phenotype. The highest scores occurred on chromosome 1p36 (*GBD2*) and 1p34 (*GBD3*), and seven other genetic locations across the genome exhibited suggestive evidence of linkage to symptomatic GBD. Some of these chromosomal regions corresponded with the locations of *Lith* genes that influence gallstone formation in inbred mouse models (Table 2).

Single-gene mutations causing haemolytic anaemias (hereditary spherocytosis, sickle-cell disease, thalassaemia, and erythrocyte enzyme deficiencies) are well-known risk factors for the formation of black pigment gallbladder stones. There are also gene defects that predispose to increased enterohepatic cycling of bilirubin, another cause of pigment stone formation[10]. Such a gene defect that is associated with black pigment stone formation is cystic fibrosis (CF) (Table 2). Gallstone prevalence in CF is 10–30% compared with <5% in age-matched controls, but biliary cholesterol saturation does not differ between CF patients with and without gallstones[11]. Experimental findings in mice with mutations in the *CFTR* gene indicate that enterohepatic cycling, as well as biliary secretion and deconjugation of bilirubin, are increased in CF. Accordingly, CF patients with gallstones more often carry the Gilbert syndrome-associated *UGT1A1* mutation compared with stone-free patients[12]. In support of this concept an increased prevalence of the Gilbert mutation has also been detected in stone patients with hereditary spherocytosis, sickle-cell anaemia and thalassaemia as compared to stone-free patients with the same conditions, all of which increase the biliary supply of unconjugated bilirubin due to intermittent haemolysis.

The characterization of gallstone formation in transgenic and knockout mice[13,14] and the genetic dissection of gallstone-susceptibility loci in inbred mice led to the identification of various *Lith* loci influencing gallstone formation in mice[15,16], but studies of the orthologous chromosomal regions in human populations have only recently been started[17]. In the future, whole-genome association studies and refined haplotype mapping in gallstone patients are likely to identify the whole set of common human *GBD* genes, enabling us to specify the individual risk for the development of GBD. This might lead to the design of new interventions, which extend our currently limited strategies for treatment and prevention of this exceptionally prevalent digestive disease.

Acknowledgements

The authors' experimental work relating to the genetics of gallstone disease has been supported by research grants from the Deutsche Forschungsgemeinschaft.

References

1. Lammert F, Sauerbruch T. Mechanisms of disease: the genetic epidemiology of gallbladder stones. Nat Clin Pract Gastroenterol Hepatol. 2005;2:423–33.
2. Völzke H, Baumeister SE, Alte D et al. Independent risk factors for gallstone formation in a region with high cholelithiasis prevalence. Digestion. 2005;71:97–105.
3. Portincasa P, Moschetta A, Palasciano G. Cholesterol gallstone disease. Lancet. 2006;368: 230–9.
4. Miquel JF, Covarrubias C, Villaroel L et al. Genetic epidemiology of cholesterol cholelithiasis among Chilean Hispanics, Amerindians, and Maoris. Gastroenterology. 1998;115:937–46.
5. Maurer KJ, Ihrig MM, Rogers AB et al. Identification of cholelithogenic enterohepatic Helicobacter species and their role in murine cholesterol gallstone formation. Gastroenterology. 2005;128:1023–33.
6. Katsika D, Grjibovski A, Einarsson C, Lammert F, Lichtenstein P, Marschall HU. Genetic and environmental influences on symptomatic gallstone disease: a Swedish study of 43 141 twin pairs. Hepatology. 2005;41:1138–43.
7. Rosmorduc O, Hermelin B, Boelle PY, Parc R, Taboury J, Poupon R. ABCB4 gene mutation-associated cholelithiasis in adults. Gastroenterology. 2003;125:452–9.

8. Lancellotti S, Zaffanello M, Di Leo E, Costa L, Lonardo A, Tarugi P. Pediatric gallstone disease in familial hypobetalipoproteinemia. J Hepatol. 2005;43:188–91.
9. Puppala S, Dodd GD, Fowler S et al. A genomewide search finds major susceptibility loci for gallbladder disease on chromosome 1 in Mexican Americans. Am J Hum Genet. 2006; 78:377–92.
10. Brink MA, Slors JF, Keulemans YC et al. Enterohepatic cycling of bilirubin: a putative mechanism for pigment gallstone formation in ileal Crohn's disease. Gastroenterology. 1999;116:1420–7.
11. Angelico M, Gandin C, Canuzzi P et al. Gallstones in cystic fibrosis: a critical reappraisal. Hepatology. 1991;14:768–75.
12. Wasmuth HE, Keppeler H, Herrmann U, Schirin-Sokhan R, Barker M, Lammert F. Coinheritance of Gilbert syndrome-associated UGT1A1 mutation increases gallstone risk in cystic fibrosis. Hepatology. 2006;43:738–41.
13. Wang DQ, Schmitz F, Kopin AS, Carey MC. Targeted disruption of the murine cholecystokinin-1 receptor promotes intestinal cholesterol absorption and susceptibility to cholesterol cholelithiasis. J Clin Invest. 2004;114:521–8.
14. Moschetta A, Bookout AL, Mangelsdorf DJ. Prevention of cholesterol gallstone disease by FXR agonists in a mouse model. Nat Med. 2004;10:1352–8.
15. Lammert F, Carey MC, Paigen B. Chromosomal organization of candidate genes involved in cholesterol gallstone formation: a murine gallstone map. Gastroenterology. 2001;120: 221–38.
16. Lyons MA, Wittenburg H. Cholesterol gallstone susceptibility loci: a mouse map, candidate gene evaluation, and guide to human *LITH* genes. Gastroenterology. 2006;131:1943–70.
17. Schafmayer C, Tepel J, Franke A et al. Investigation of the *Lith1* candidate genes *ABCB11* and *LXRA* in human gallstone disease. Hepatology. 2006;44:650–65.
18. Wittenburg H, Lammert F. Genetic predisposition to gallbladder stones. Semin Liver Dis. 2007;27:109–21.
19. Van Mil SW, van der Woerd WL, van der Brugge G et al. Benign recurrent intrahepatic cholestasis type 2 is caused by mutations in ABCB11. Gastroenterology. 2004;127:379–84.
20. Miller LJ, Holicky EL, Ulrich CD, Wieben ED. Abnormal processing of the human cholecystokinin receptor gene in association with gallstones and obesity. Gastroenterology. 1995;109:1375–80.
21. Pullinger CR, Eng C, Salen G et al. Human cholesterol 7alpha-hydroxylase (CYP7A1) deficiency has a hypercholesterolemic phenotype. J Clin Invest. 2002;110:109–17.
22. Jiang ZY, Han TQ, Suo GJ et al. Polymorphisms at cholesterol 7alpha-hydroxylase, apolipoproteins B and E and low density lipoprotein receptor genes in patients with gallbladder stone disease. World J Gastroenterol. 2004;10:1508–12.
23. Juvonen T, Savolainen MJ, Kairaluoma MI, Lajunen LH, Humphries SE, Kesäniemi YA. Polymorphism at the apoB, apoA-I, and the cholesterol ester transfer protein gene loci in patients with gallbladder disease. J Lipid Res. 1995;36:804–12
24. Dixit M, Choudhuri G, Mittal B. Association of lipoprotein receptor, receptor-associated protein, and metabolizing enzyme gene polymorphisms with gallstone disease: case–control study. Hepatol Res. 2006;36:61–9.

4
Genetics in viral hepatitis: role of HLA class I and II alleles in hepatitis C virus infection

R. THIMME and C. NEUMANN-HAEFELIN

INTRODUCTION

Hepatitis C virus (HCV) is a small single-stranded RNA virus that belongs to the Flaviviridae virus family. More than 120 million people worldwide are chronically HCV-infected[1]. The factors that determine the outcome and natural course of infection are not completely understood. However, it is generally accepted that host immune responses to HCV, especially T cell responses, play an important role in control of HCV infection as well as disease pathogenesis[2]. Indeed, viral clearance during self-limited HCV infection is characterized by a vigorous polyclonal $CD4^+$ (refs 3–5) and $CD8^+$ (refs 6 and 7) T cell response. These responses first become detectable in the peripheral blood, but accumulate rapidly in the liver where they cause liver disease and viral clearance[8,9]. The central role of $CD8^+$ T cells has also been demonstrated in studies showing a clear temporal association between the emergence of virus-specific $CD8^+$ T cells and the first rapid decline in viraemia[10,11]. A critical role for both T cell subsets in HCV clearance was recently directly proven in chimpanzees. Importantly, antibody-mediated depletion of $CD8^+$ cells prolonged viraemia after rechallenge of previously immune animals with the same viral strain, and viral clearance precisely correlated with recovery of HCV-specific $CD8^+$ T cells within the liver[12]. The depletion of $CD4^+$ cells in a parallel experiment resulted in HCV persistence, supporting the important role of memory $CD4^+$ T cells for the outcome of HCV infection[13].

Importantly, several groups have shown an association between certain human leucocyte antigen (HLA) alleles and the outcome of infection. The major histocompatibility complex (MHC) plays a critical role in the immune response, since these gene products present viral antigen to CD8 (class I) and CD4 (class II) T cells (Figure 1). Class I and II HLA are encoded by very polymorphic genes. Polymorphisms in the peptide-binding regions of these molecules determine antigen specificities and the strength of the immune response. Possible differences in presentation might influence the strength of

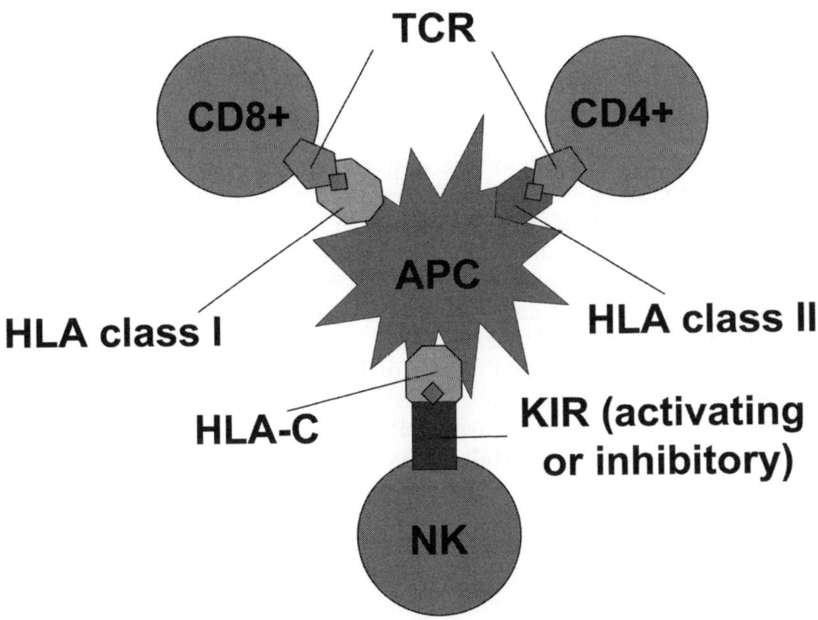

Figure 1 Interaction of $CD8^+$ T cells, $CD4^+$ T cells and NK cells with HLA molecules. $CD8^+$ T cells recognize peptides specific for their T cell receptor (TCR) which are presented by antigen-presenting cells (APC) via HLA class I molecules. $CD4^+$ T cells recognize peptides presented via HLA class II alleles. NK cells receive regulatory (inhibitory or activating) signals via binding of HLA class I alleles to killer cell immunoglobulin-like receptors (KIR)

the T cell response and breadth of the T cell receptor (TCR) repertoire, and thus the ease with which the virus can escape or be eliminated. Of note, several studies performed in well-defined cohorts have suggested that class I and II alleles may also play an important role in the outcome of HCV infection. Indeed, studies of women who were accidentally infected with the same viral strain (HCV contaminated immunoglobulin D antibody) showed that viral diversity could not determine clearance versus persistence[14–18]. Thus, differences in the host response through genetic restriction of the virus-specific T cell response were more likely to be the determining factor.

ROLE OF CLASS I ALLELES IN HCV INFECTION

The prominent role of $CD8^+$ cells in HCV infection has led to the hypothesis that the genetic restriction of the $CD8^+$ T cell response may play an important role in the outcome of infection. To test this hypothesis Thio et al. analysed three distinct cohorts of individuals, including 231 persons with well-documented clearance of an HCV infection and 444 matched persistently infected persons[19]. Of note, they observed that HLA-A*1101 (odds ratio (OR) 0.49; 95% CI 0.27–0.89), B*57 (OR 0.62; 95% CI 0.39–1) and Cw*0102 (OR 0.43; 95% CI 0.21–0.89) were associated with viral clearance and that HLA-A*2301 (OR 1.78; 95% CI 1.01–3.11) and Cw*04 (OR 1.78; 95% CI 1.21–2.59) were associated with viral persistence. The association of HLA-Cw*04 with HCV persistence was codominant (two copies of the gene were associated more strongly with persistence than one copy). Despite the association between HLA-Cw*04 homozygosity and HCV persistence, this study did not detect an overall association between class I homozygosity and viral persistence. This was surprising because one would suspect that individuals who are fully heterozygous for class I alleles can present a more diverse array of epitopes to virus-specific $CD8^+$ T cells and thus are more likely to clear the virus.

A study performed by McKiernan et al. examined the association between class I alleles and outcome in a cohort of Irish females inoculated with HCV genotype 1b from a single source[17]. Class I typing was performed in a total of 227 female subjects, 141 with chronic infection and 86 with spontaneous viral clearance. The study was undertaken 20 years following inoculation with HCV and only females with positive anti-HCV antibodies were included. Of note, A*023, B*27 and Cw*01 occurred more frequently in individuals with viral clearance (39.5%, 14% and 9.3%, respectively) compared to patients with chronic infection (19.1%, 2.1% and 1.4%, respectively, $p<0.005$). B*08 occurred more often in those with chronic infection compared to those who cleared the virus (39.7% vs 19.8%, $p = 0.002$). Thus, the results obtained in this particular well-defined cohort differ somewhat from the results obtained by Thio et al.[19] in a study population whose baseline characteristics were more heterogeneous in terms of race, sex, coinfection with other viruses and viral genotype. The resulting variance in antigen specificities may account for some differences between these two studies.

Of note, in the Irish cohort the HLA allele with the strongest association with viral clearance was HLA-B*27. Thus, together with HLA-B*57, a picture emerges that the same HLA alleles may confer protection in HCV and HIV infection. The immunological basis for the significant association between HLA-B*27 and HCV clearance has previously been unknown since, in contrast to HIV, no B*27-restricted epitopes had been identified. In this context we recently described an HLA-B*27-restricted epitope within the viral NS5B protein that is recognized in most HLA-B*27-positive women from the Irish cohort who spontaneously cleared the virus[20]. Indeed, five of six women tested responded to this viral epitope, and it was the dominant response in all women. Of note, such a clear dominance of a single epitope-specific $CD8^+$ T cell response has not been shown for any other HLA allele. Thus, these results suggest that a single HLA-B*27-restricted $CD8^+$ T cell response may explain the observed association between spontaneous viral clearance and HLA-B*27 (Figure 2).

The biological role of the NS5B-specific $CD8^+$ T cell response was further supported by the finding that almost all chronically HCV-infected patients expressing HLA-B*27 allele demonstrated sequence polymorphisms within this epitope[20]. The frequent detection of sequence variations within HLA-B*27-positive patients was in clear contrast to HLA-B*27-negative patients

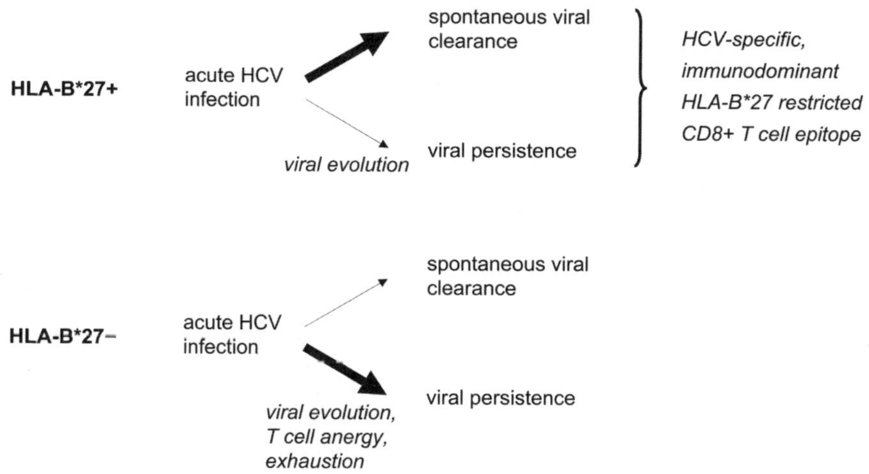

Figure 2 Impact of HLA-B*27 on the course of HCV infection. The course of acute HCV infection in HLA-B*27-positive (upper panel) and -negative (lower panel) patients is shown. The majority of HLA-B*27-positive patients is able to clear the virus spontaneously; a minority progresses to viral persistence, which is then associated with viral evolution. Viral clearance as well as viral evolution can both be mechanistically linked to a single HCV-specific immunodominant HLA-B*27-restricted $CD8^+$ T cell epitope. In the majority of HLA-B*27-negative patients HCV persists; different mechanisms of T cell failure, e.g. viral evolution, T cell anergy, and T cell exhaustion, are involved in viral persistence

who did not show viral escape at this epitope, clearly suggesting that selective pressure leading to escape was associated with the HLA-B*27 allele. Taken together, these results suggest that a single HLA-B*27-restricted $CD8^+$ T cell response may explain the association between HLA-B*27 and spontaneous viral clearance as well as HLA-B*27 and sequence polymorphisms within the NS5B protein. These results support the notion that viral escape mutations, at least within immunodominant epitopes restricted by specific HLA genes, contribute to viral persistence, as has been suggested by other investigators[21]. For example, Timm et al. recently observed a strong association between sequence variation within an immunodominant HLA-B*8-restricted NS3 epitope and expression of HLA-B*8, also supporting reproducible allele-specific selection pressures at the population level[22]. Ray et al. performed a study in the well-documented Irish cohort and compared sequences present in serum obtained 18–22 years after infection to sequences present in the infecting inoculum[23]. Interestingly, they observed that amino acid substitutions in known epitopes were directed away from consensus in women expressing the restricting HLA allele (suggesting immune selection) and towards consensus in those lacking the allele (suggesting reversion). For example, sequence changes in B*35-restricted epitopes were observed only in women carrying that allele, and sequence changes in a HLA-B*37-restricted epitope were 8.5 times as likely to occur in women expressing the allele as compared with those who lacked it ($p<0.001$). In contrast, sequence changes in HLA-A*01- and B*08-restricted epitopes occurred more frequently in subjects who did not express these alleles, and represented changes towards the consensus residue and the absence of $CD8^+$ T cell pressure. In addition, HCV adaptation to HLA class I-restricted immune pressure was recently shown in two host populations from West Australia and Switzerland[24]. Thus, these combined results also show adaptation to multiple HLA alleles across multiple genes in HCV.

It is still unclear why most HCV-infected HLA-B*27-positive patients clear the virus and why only a few show viral persistence that seems to be associated with the emergence of viral escape mutations. It is possible that viral factors and/or certain characteristics of the adaptive immune response contribute to the development of viral escape mutations. For example, it has been hypothesized that a major factor underlying the apparently less variable nature of certain regions of the genome of mutable RNA viruses is viral fitness cost, e.g. mutations within certain regions will lead to loss of protein structures or enzymatic functions[25]. In this regard it is interesting to note that the NS5B region targeted by the immunodominant HLA-B*27 epitope is conserved in the general population of HLA-B*27-negative patients. Thus, it is possible that mutations within this dominant epitope do not arise rapidly, thereby allowing a sustained antiviral $CD8^+$ T cell response required for spontaneous viral clearance in most infected patients. It is also possible that immunological factors, such as TCR diversity[26] or a lack of a sufficient $CD4^+$ T cell help contribute to escape from $CD8^+$ T cell responses. The latter has been elegantly shown in chimpanzees, where an incomplete control of HCV replication by memory $CD8^+$ T cells occurred in the absence of adequate $CD4^+$ T cell help and was associated with the emergence of viral escape mutations[13]. Clearly, additional studies are needed to address these important issues.

In contrast to the significant role of HLA class I alleles for the outcome of infection in acute HCV infection, it was recently reported that HLA class I diversity has only a minor influence on fibrosis progression or disease severity in chronic infection. The study population consisted of 670 adults with chronic HCV infection, including 155 with advanced cirrhosis and 237 non-HCV controls[27]. These results suggest that hepatic injury is more likely to be determined by a complex interplay between host, viral and genetic factors. They also suggest that genetic restriction may play an important role in the early phase of infection determining the outcome. Once viral persistence is established, however, specific genetic restrictions of virus-specific $CD8^+$ T cell responses may not be of major relevance. This interesting finding can easily be explained by the early occurrence of viral escape mutations or other mechanisms that down-tune virus-specific T cell responses in chronic infection.

Taken together, these results suggest a scenario in which the genetic restriction of the virus-specific $CD8^+$ T cell response may be a key dominant determinant of the virus-specific $CD8^+$ T cell response and thus the outcome of infection during the acute phase of infection but not during chronic infection. However, these results also show that other virological (e.g. viral fitness), immunological (e.g. regulatory T cells) or genetic factors (e.g. genetic restriction of $CD4^+$ T cell responses) may significantly contribute to the outcome of HCV infection. A better understanding of these dynamic and complex processes will require the analysis of the host–virus interaction in larger and better-defined patient cohorts, or in novel cell culture or small animal models.

ROLE OF CLASS I AND KIR COMBINATIONS IN HCV INFECTION

It has been recently described that killer cell immunoglobulin-like receptors (KIR) on natural killer (NK) cells bind to HLA class I molecules, thus transducing activating or inhibitory signals. Therefore, protective effects of HLA class I alleles might not only be mediated by $CD8^+$ T cells, but might also involve NK cells.

NK cells are an important component of the innate immune system, secreting antiviral cytokines and lysing infected cells at an early stage of viral infections. They are regulated by different activating and inhibitory signals, including the interaction between activating or inhibitory KIR and HLA class I molecules on target cells. KIR are transmembrane proteins on NK cells, containing two or three immunoglobulin-like domains and a long or short cytoplasmic tail (2DL, 3DL, 2DS, 3DS). Some KIR have defined HLA class I ligands. The KIR 2DL2/2DL3 and 3DL/3DS1 segregate as alleles[28]. The KIR associated with different outcomes of HCV infection are KIR2DL1, 2, and 3[29–31] (Figure 3). KIR2DL1 is present in almost all individuals and strongly binds HLA-C alleles that harbour a lysine at position 80 (HLA-C2 group). This interaction leads to a strong inhibitory signal on NK cells. KIR2DL2 and 3, by contrast, segregate as alleles and bind HLA-C alleles with an asparagine at position 80 (HLA-C1 group). However, KIR2DL3 binds only weakly, and thus results in weak inhibition of NK cells only. Accordingly, individuals homozygous for KIR2DL3 and HLA-C1 alleles have little inhibitory signals. Homozygosity for KIR2DL3 implicates lack of the more potent HLA-C1 ligand KIR2DL2, and homozygosity for HLA-C1 alleles implicates lack of HLA-C2 alleles which are strong binders to the more generally expressed and highly inhibitory KIR2DL1. According to this model, homozygosity for KIR2DL3 and HLA-C1 alleles should result in a reduced threshold of NK cell activation and might therefore be associated with a favourable outcome of HCV infection. Indeed, individuals homozygous for KIR2DL3 and HLA-C1 alleles have been reported to have a greater chance of eliminating HCV: the frequency of the homozygous genotype was 20.4% in individuals who eliminated HCV infection, but only 9.9% in individuals with persistent infection after needle-stick infection (OR 2.33; $p = 0.001$)[29]. This protective effect was not found in individuals who obtained inocula with high viral titres (e.g. HCV-contaminated blood transfusions); this might be due to the inhibition of NK cell function by viral factors, such as the HCV E2 glycoprotein[32,33].

In sum, HLA class I alleles can contribute to outcome of HCV infection by mechanisms distinct from restriction of $CD8^+$ T cell responses. However, it seems unlikely that the interaction with KIR on NK cells can account for the strong association of certain HLA class I alleles with HCV clearance or persistence[34].

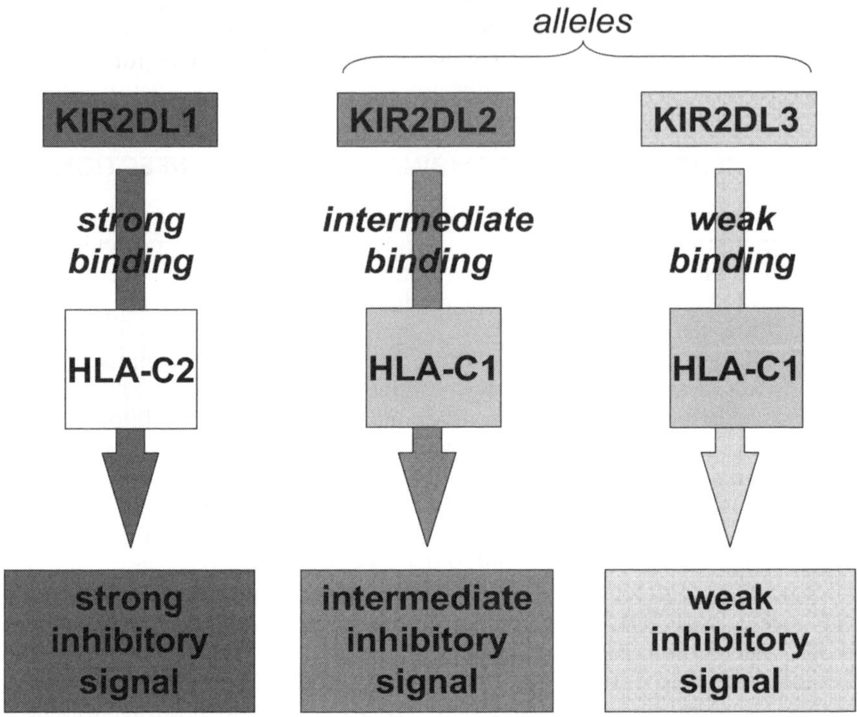

Figure 3 Interaction of KIR2DL1, 2, and 3 with HLA class I molecules. KIR2DL1 binds strongly to HLA-C group 2. This results in a strong inhibitory signal. KIR2DL2 and 3 function as alleles. KIR2DL2 shows intermediate binding to HLA-C group 1, while KIR2DL3 binds only weakly to HLA-C group 1. This results in intermediate and weak inhibitory signals on NK cell function, respectively.

Role of class II alleles in HCV infection

While virus-specific $CD8^+$ T cells are thought to be the major effector cells in viral infections, virus-specific $CD4^+$ T cells have important helper functions, e. g. augmentation of humoral responses and $CD8^+$ T cell responses. In addition, they also have a direct antiviral effect by secreting antiviral cytokines. Of note, an association between vigorous HCV-specific $CD4^+$ T cell responses and resolution of acute HCV infection was demonstrated more than a decade ago[3,5]. Different methodical approaches, including proliferation assays using recombinant proteins or synthetic peptides[3,5], *ex-vivo* intracellular cytokine staining with peptides[35,36] and *ex-vivo* multimer staining[37] demonstrated that spontaneous HCV clearance is associated with strong and multispecific $CD4^+$ T cell responses, which are sustained after viral clearance. Most studies found that, in resolving acute HCV infection, $CD4^+$ T cell responses target mainly non-structural proteins, while in persistent HCV infection, $CD4^+$ T cell responses to the core protein predominate[3,4,38–40]. The chimpanzee model of HCV allowed a more direct demonstration of the effect of CD4$^+$ T cells: the antibody-mediated depletion of CD4$^+$ cells resulted in prolonged viraemia, emergence of viral quasispecies containing escape mutations within targeted CD8$^+$ T cell epitopes, and finally viral persistence[13]. Since CD4$^+$ T cell responses are restricted by HLA class II molecules (Figure 1), it has been postulated that, similar to HLA class I alleles, certain HLA class II alleles might be associated with different outcomes of HCV infection. Indeed, approximately 50 studies have been carried out to date addressing the association between HLA class II alleles and the outcome of HCV infection. These studies were heterogeneous with respect to different characteristics of the study population (ethnicity, resolved infection versus chronic infection, healthy controls versus chronic infection, sex, age, duration of infection, mode of infection, HCV genotype). Nevertheless, an impressive number of studies showed an association between HCV clearance and the HLA class II alleles DRB1*1101 and DQB1*0301, which are genetically closely linked, a phenomenon referred to as linkage disequilibrium. The studies were conducted in different countries, including European, Asian, and North American people (Table 1). In two independent meta-analyses the association between DRB1*1101 and viral clearance had an OR of 2.5 (95% CI 1.7–3.7) and 2.02 (95% CI 1.6–2.6), respectively, while the association between DQB1*03 and viral clearance had an OR of 3.0 (95% CI 1.8–4.8) and 2.4 (95% CI 1.6–3.4), respectively[41,42]. Notably, different factors may have contributed to an overinterpretation of the protective role of DRB1*11 and DQB1*0301 in both meta-analyses. First, there might be a bias towards the publication of studies with positive HLA associations, while studies showing no significant association with any HLA allele are less likely to be published[43]. Second, the meta-analyses could include only studies with an OR for DRB1*11 and/or DQB1*0301; studies which identified positive associations for other HLA alleles but did not state an OR for DRB1*11 and/or DQB1*0301 were not included.

In the well-defined Irish cohort, DRB1*01 was associated with spontaneous HCV clearance as in all studies published[14–18]. DRB1*01 was expressed in 22–32% of all women with viral clearance and in 7–9% of women with chronic infection ($p < 0.05$ in all studies; OR 2.7–5.2). In addition, the study by McKiernan et al.[17] also identified DRB1*0401 and DRB1*15 as protective alleles, a finding which is also supported by two British and German studies[44,45]. Strikingly, DRB1*1101 was not found to be protective in the Irish cohort by either of the studies. This marked difference between different ethnic groups, the protective role of DRB1*11 in continental Europeans but not in Irish subjects, and the protective role of DRB1*01 in Irish but not continental European subjects, is further underlined by the finding that DRB1*0101 is protective in white North Americans but not Afro-Americans, while DQB1*0301 (which is in linkage disequilibrium with DRB1*1101) is protective in Afro-Americans, but not white North Americans[46].

In summary, there is strong evidence for a protective role of the HLA class II alleles DRB1*1101/ DQB1*0301 and DRB1*0101, depending on the ethnic background (and/or other factors as HCV genotype), and some evidence for a

Table 1 HLA class I and II alleles associated with HCV spontaneous clearance

	Country/ethnic group	*Reference*
HLA class I		
As, B27	Irish	17
A11*01, B*57, Cw*0102	US	19
HLA class II		
DRB1*0101	Irish	14–18
	US (White subgroup)	46
DRB1*04(01)	UK	45
	Irish	17
DRB1*11(01)	Italian	54[a]
	Turkish	55
	UK	56, 57
	Italian	58[b], 59[b]
	French	60
DRB1*15	Irish	17
	German	44[c]
DQB1*0301	US (Black subgroup)	46
	Polish	61
	UK	45, 56, 57
	Italian	58, 59
	French	60
	Northern European	62[d]
DQB1*0501	Irish	18
	US (White subgroup)	46

[a]DR11, [b]DRB1*1104, [c]DRB1*15011, [d]DQB1*0302.

protective role of DRB1*04 and DRB1*15 (compare Table 1). In addition, several studies have shown correlations between certain HLA class II alleles and progression of liver disease, ALT levels, viral load, and response to antiviral therapy in chronic HCV infection (reviewed in ref. 41).

There are several difficulties in the analysis of the mechanism of protection by these HLA class II alleles due to the characteristics of $CD4^+$ T cell epitopes. Many $CD4^+$ T cell epitopes show a highly degenerate binding to multiple HLA class II alleles and are promiscuously recognized in patients with different HLA class II alleles. In addition, the minimal stimulating sequences (usually 8–13 amino acid residues) and the optimal stimulating sequences (usually 10–15 amino acid residues) can differ substantially. Thus, the HLA restriction as well as minimal and optimal stimulating sequence may vary for different T cell clones intra- or inter-individually, although they are specific for the same epitope[47]. Due to these difficulties, little is known about the HLA restriction and fine specificity of many HCV-specific $CD4^+$ T cell responses. In addition, in peripheral blood HCV-specific $CD4^+$ T cells are detectable only at low frequencies compared to HCV-specific $CD8^+$ T cells, even in the acute phase of infection, and HLA class II multimers have only been introduced recently[37].

The production of antigen-specific $CD4^+$ T cell lines and clones allowed the fine mapping and determination of HLA restriction of several $CD4^+$ T cell epitopes in resolving acute HCV infection[48,49]. Some of these epitopes have been shown to be immunodominant, e.g. were targeted by the majority of generated $CD4^+$ T cell clones from nearly all patients with resolving acute HCV infection, and their sequence was highly conserved in and between genotypes. Strikingly, these $CD4^+$ T cell epitopes were restricted predominantly by HLA class II alleles DRB1*11, DRB1*04, DRB1*15 and DQB1*0301, which have been shown to have a protective role. Furthermore, these peptides bound to the majority of relevant DRB1 molecules in an *in-vitro* binding assay. Interestingly, for all but one peptide binding was superior to the DRB1*0101 allele ($IC_{50} \leqslant 10$ nM; $IC_{50} \leqslant 500$ nM indicating binding), which has also been reported to be protective in the Irish cohort. One single immunodominant epitope, NS3 1251–1259 (minimal epitope as described in ref. 48), has been analysed in more detail by several authors. Of note, most studies detected responses to this epitope in a majority of patients with resolving acute HCV infection[4,40,48,49], while a disappearance of this $CD4^+$-specific T cell response was associated with a loss of viral control in one patient[48]. This epitope is highly conserved in all HCV genotypes and it has been shown to be restricted by HLA class II alleles DRB1*0401, *1001, *1101, *1201, *1302, *1601, DQB1*0301, and DRB3*0301 in T cell clones[40,48,49]. HLA class II *in-vivo* binding assays demonstrated binding to DRB1*0101, *0401, *0404, *0405, *1101, *1201, *1302, *0701, *0802, *0901, and *1501 and multimers containing DRB1*0401 and *1101, respectively, have been shown to stain significant numbers of $CD4^+$ T cells in resolving/resolved acute HCV infection[37,50]. These results suggest that this single epitope may dominate the $CD4^+$ T cell response in resolving acute HCV infection and may thus lead to viral clearance[48]. However, other studies reported NS3 1251-specific $CD4^+$ T cell responses only in a minority of patients[51,52]. Further, even in studies where the epitope is recognized by most patients, it is not dominant (e.g. the strongest response) in any individual[50,52].

More recently a comprehensive analysis of the HCV-specific $CD4^+$ T cell response was performed using overlapping peptides covering the NS3/NS4 proteins[47] and the complete HCV protein, respectively[36,40]. These studies identified 10 and six peptides, respectively, which were each recognized by $\geqslant 30\%$ of patients with resolving acute HCV infection. In the latter study all subjects spontaneously resolving the HCV infection mounted an immune response against at least one of these epitopes. Interestingly, the specificity as well as HLA restriction in the NS3/NS4 region overlapped significantly between the two studies. The majority of immunodominant epitopes were shown to be restricted by HLA class II alleles DRB1*04, *11, and *15. In agreement with previous studies, however, the immunodominant epitopes showed degenerate HLA binding in an *in-vitro* binding assay. In the study analysing the complete HCV polyprotein[40], all individuals with resolving acute HCV infection targeted at least one of the six immunodominant epitopes, and these were highly conserved in the sequences published. In total, individuals targeted 1–28 $CD4^+$ T cell epitopes.

The fact that $CD4^+$ T cell epitopes which are recognized in most patients are highly conserved may be due to an experimental artifact. Patients were analysed with overlapping peptides derived from genotype 1a, while the genotypes of the infecting viruses were diverse. Thus, responses against conserved epitopes might be detected more frequently in non-genotype 1a patients than less conserved epitopes. However, the number of epitopes recognized in genotype 1a and non-1a patients did not differ significantly[40]. The finding is therefore more likely to be of biological significance than just a simple artifact. It could well be that $CD4^+$ T cell responses targeting highly conserved epitopes have a protective effect, since the infecting virus cannot undergo escape mutations easily within these regions due to functional and/or structural constraints, a phenomenon referred to as viral fitness cost.

In sum, it is tempting to postulate that protective HLA class II alleles such as DRB1*11 (and probably DRB1*01 in the Irish cohort) restrict a range of immunodominant epitopes that target highly conserved epitopes and are thus highly effective in performing helper functions and/or producing antiviral agents themselves. However, important issues concerning the determinants of protective $CD4^+$ T cell responses still remain elusive. Are there single protective epitopes, or is a broad repertoire of responses in an individual essential for viral clearance? What is the role of the kinetics of viral spread and the immune response? What is the impact of $CD4^+$ T cell phenotype on the outcome of infection[53]? These and other questions need to be addressed in future studies.

CONCLUSIONS

Growing evidence suggests a central role of the genetic restriction of virus-specific T cell responses in different viral infections. Interestingly, the same HLA alleles, such as B*27, have a protective role in different viral infections, such as HIV or HCV, that is mediated by specific $CD8^+$ T cell responses. Immunological factors that may explain this strong association include among others a high viral peptide-binding affinity. However, it is also important to note that virological factors, such as the ability of viral regions to tolerate escape mutations (viral fitness cost), have a strong impact on the efficiency of the virus-specific $CD8^+$ T cell response. Thus, the outcome of HCV infection is determined by a complex interplay of host genetic, immunological and virological factors. Better insights into these host–virus interactions may lead to better insights into the mechanisms of HCV clearance and persistence, but also into the development of new antiviral and vaccination strategies.

Acknowledgements

The authors thank all members of our laboratory for their excellent contributions, as well as Prof. Dr. Dr. h. c. mult. H.E. Blum and Dr. Jörg Timm for helpful discussions and critically reading the manuscript. C.N.-H. is supported by the Kompetenznetz Hepatitis (HepNet, BMWF), R.T. is supported by the Deutsche Forschungsgemeinschaft (TH 719/2-2 and 2-3, Emmy Noether Programm and SFB 620).

References

1. Shepard CW, Finelli L, Alter MJ. Global epidemiology of hepatitis C virus infection. Lancet Infect Dis. 2005;5:558–67.
2. Dustin LB, Rice CM. Flying under the radar: the immunobiology of hepatitis C. Annu Rev Immunol. 2007;25:71–99.
3. Diepolder HM, Zachoval R, Hoffmann RM et al. Possible mechanism involving T-lymphocyte response to non-structural protein 3 in viral clearance in acute hepatitis C virus infection. Lancet. 1995;346:1006–7.
4. Gerlach JT, Diepolder HM, Jung MC et al. Recurrence of hepatitis C virus after loss of virus-specific CD4(+) T-cell response in acute hepatitis C. Gastroenterology. 1999;117:933–41.
5. Missale G, Bertoni R, Lamonaca V et al. Different clinical behaviors of acute hepatitis C virus infection are associated with different vigor of the anti-viral cell-mediated immune response. J Clin Invest. 1996;98:706–14.
6. Gruener NH, Gerlach TJ, Jung MC et al. Association of hepatitis C virus-specific CD8+ T cells with viral clearance in acute hepatitis C. J Infect Dis. 2000;181:1528–36.
7. Lechner F, Wong DK, Dunbar PR et al. Analysis of successful immune responses in persons infected with hepatitis C virus. J Exp Med. 2000;191:1499–512.
8. Thimme R, Bukh J, Spangenberg HC et al. Viral and immunological determinants of hepatitis C virus clearance, persistence, and disease. Proc Natl Acad Sci USA. 2002;99:15661–8.
9. Cooper S, Erickson AL, Adams EJ et al. Analysis of a successful immune response against hepatitis C virus. Immunity. 1999;10:439–49.
10. Thimme R, Oldach D, Chang KM, Steiger C, Ray SC, Chisari FV. Determinants of viral clearance and persistence during acute hepatitis C virus infection. J Exp Med. 2001;194:1395–406.
11. Cox AL, Mosbruger T, Lauer GM, Pardoll D, Thomas DL, Ray SC. Comprehensive analyses of CD8+ T cell responses during longitudinal study of acute human hepatitis C. Hepatology. 2005;42:104–12.
12. Shoukry NH, Grakoui A, Houghton M et al. Memory CD8+ T cells are required for protection from persistent hepatitis C virus infection. J Exp Med. 2003;197:1645–55.
13. Grakoui A, Shoukry NH, Woollard DJ et al. HCV persistence and immune evasion in the absence of memory T cell help. Science. 2003;302:659–62.
14. Fanning LJ, Levis J, Kenny-Walsh E, Wynne F, Whelton M, Shanahan F. Viral clearance in hepatitis C (1b) infection: relationship with human leukocyte antigen class II in a homogeneous population. Hepatology. 2000;31:1334–7.
15. Barrett S, Goh J, Coughlan B et al. The natural course of hepatitis C virus infection after 22 years in a unique homogeneous cohort: spontaneous viral clearance and chronic HCV infection. Gut. 2001;49:423–30.
16. Barrett S, Ryan E, Crowe J. Association of the HLA-DRB1*01 allele with spontaneous viral clearance in an Irish cohort infected with hepatitis C virus via contaminated anti-D immunoglobulin. J Hepatol. 1999;30:979–83.
17. McKiernan SM, Hagan R, Curry M et al. Distinct MHC class I and II alleles are associated with hepatitis C viral clearance, originating from a single source. Hepatology. 2004;40:108–14.
18. McKiernan SM, Hagan R, Curry M et al. The MHC is a major determinant of viral status, but not fibrotic stage, in individuals infected with hepatitis C. Gastroenterology. 2000;118:1124–30.
19. Thio CL, Gao X, Goedert JJ et al. HLA-Cw*04 and hepatitis C virus persistence. J Virol. 2002;76:4792–7.
20. Neumann-Haefelin C, McKiernan S, Ward S et al. Dominant influence of an HLA-B27 restricted CD8+ T cell response in mediating HCV clearance and evolution. Hepatology. 2006;43:563–72.
21. Bowen DG, Walker CM. Mutational escape from CD8+ T cell immunity: HCV evolution, from chimpanzees to man. J Exp Med. 2005;201:1709–14.
22. Timm J, Lauer GM, Kavanagh DG et al. CD8 epitope escape and reversion in acute HCV infection. J Exp Med. 2004;200:1593–604.

23. Ray SC, Fanning L, Wang XH, Netski DM, Kenny-Walsh E, Thomas DL. Divergent and convergent evolution after a common-source outbreak of hepatitis C virus. J Exp Med. 2005;201:1753–9.
24. Gaudieri S, Rauch A, Park LP et al. Evidence of viral adaptation to HLA class I-restricted immune pressure in chronic hepatitis C virus infection. J Virol. 2006;80:11094–104.
25. Altman JD, Feinberg MB. HIV escape: there and back again. Nat Med. 2004;10:229–30.
26. Meyer-Olson D, Shoukry NH, Brady KW et al. Limited T cell receptor diversity of HCV-specific T cell responses is associated with CTL escape. J Exp Med. 2004;200:307–19.
27. Patel K, Norris S, Lebeck L et al. HLA class I allelic diversity and progression of fibrosis in patients with chronic hepatitis C. Hepatology. 2006;43:241–9.
28. Williams AP, Bateman AR, Khakoo SI. Hanging in the balance. KIR and their role in disease. Mol Interv. 2005;5:226–40.
29. Khakoo SI, Thio CL, Martin MP et al. HLA and NK cell inhibitory receptor genes in resolving hepatitis C virus infection. Science. 2004;305:872–4.
30. Parham P. Immunology. NK cells lose their inhibition. Science. 2004;305:786–7.
31. Rajagopalan S, Long EO. Understanding how combinations of HLA and KIR genes influence disease. J Exp Med. 2005;201:1025–9.
32. Tseng CT, Klimpel GR. Binding of the hepatitis C virus envelope protein e2 to CD81 inhibits natural killer cell functions. J Exp Med. 2002;195:43–50.
33. Crotta S, Stilla A, Wack A et al. Inhibition of natural killer cells through engagement of CD81 by the major hepatitis C virus envelope protein. J Exp Med. 2002;195:35–42.
34. Gaudieri S, DeSantis D, McKinnon E et al. Killer immunoglobulin-like receptors and HLA act both independently and synergistically to modify HIV disease progression. Genes Immun. 2005;6:683–90.
35. Godkin A, Jeanguet N, Thursz M, Openshaw P, Thomas H. Characterization of novel HLA-DR11-restricted HCV epitopes reveals both qualitative and quantitative differences in HCV-specific CD4+ T cell responses in chronically infected and non-viremic patients. Eur J Immunol. 2001;31:1438–46.
36. Day CL, Lauer GM, Robbins GK et al. Broad specificity of virus-specific CD4+ T-helper-cell responses in resolved hepatitis C virus infection. J Virol. 2002;76:12584–95.
37. Day CL, Seth NP, Lucas M et al. *Ex vivo* analysis of human memory CD4 T cells specific for hepatitis C virus using MHC class II tetramers. J Clin Invest. 2003;112:831–42.
38. Chang KM, Thimme R, Melpolder JJ et al. Differential CD4 and CD8 T-cell responsiveness in hepatitis C virus infection. Hepatology. 2001;33:267–76.
39. Rosen HR, Miner C, Sasaki AW et al. Frequencies of HCV-specific effector CD4+ T cells by flow cytometry: correlation with clinical disease stages. Hepatology. 2002;35:190–8.
40. Schulze zur Wiesch J, Lauer GM, Day CL et al. Broad repertoire of the CD4+ Th cell response in spontaneously controlled hepatitis C virus infection includes dominant and highly promiscuous epitopes. J Immunol. 2005;175:3603–13.
41. Yee LJ. Host genetic determinants in hepatitis C virus infection. Genes Immun. 2004;5: 237–45.
42. Hong X, Yu RB, Sun NX, Wang B, Xu YC, Wu GL. Human leukocyte antigen class II DQB1*0301, DRB1*1101 alleles and spontaneous clearance of hepatitis C virus infection: a meta-analysis. World J Gastroenterol. 2005;11:7302–7.
43. Azocar J, Clavijo OP, Yunis EJ. MHC class II genes in HCV viral clearance of hepatitis C infected Hispanic patients. Hum Immunol. 2003;64:99–102.
44. Lechmann M, Schneider EM, Giers G et al. Increased frequency of the HLA-DR15 (B1*15011) allele in German patients with self-limited hepatitis C virus infection. Eur J Clin Invest. 1999;29:337–43.
45. Cramp ME, Carucci P, Underhill J, Naoumov NV, Williams R, Donaldson PT. Association between HLA class II genotype and spontaneous clearance of hepatitis C viraemia. J Hepat. 1998;29:207–13.
46. Thio CL, Thomas DL, Goedert JJ et al. Racial differences in HLA class II associations with hepatitis C virus outcomes. J Infect Dis. 2001;184:16–21.
47. Gerlach JT, Ulsenheimer A, Gruner NH et al. Minimal T-cell-stimulatory sequences and spectrum of HLA restriction of immunodominant CD4+ T-cell epitopes within hepatitis C virus NS3 and NS4 proteins. J Virol. 2005;79:12425–33.

48. Diepolder HM, Gerlach JT, Zachoval R et al. Immunodominant CD4+ T-cell epitope within nonstructural protein 3 in acute hepatitis C virus infection. J Virol. 1997;71:6011–19.
49. Lamonaca V, Missale G, Urbani S et al. Conserved hepatitis C virus sequences are highly immunogenic for CD4(+) T cells: implications for vaccine development. Hepatology. 1999; 30:1088–98.
50. Harcourt GC, Lucas M, Sheridan I, Barnes E, Phillips R, Klenerman P. Longitudinal mapping of protective CD4+ T cell responses against HCV: analysis of fluctuating dominant and subdominant HLA-DR11 restricted epitopes. J Viral Hepat. 2004;11:324–31.
51. Harcourt G, Hellier S, Bunce M et al. Effect of HLA class II genotype on T helper lymphocyte responses and viral control in hepatitis C virus infection. J Viral Hepat. 2001; 8:174–9.
52. Wertheimer AM, Miner C, Lewinsohn DM, Sasaki AW, Kaufman E, Rosen HR. Novel CD4+ and CD8+ T-cell determinants within the NS3 protein in subjects with spontaneously resolved HCV infection. Hepatology. 2003;37:577–89.
53. Ulsenheimer A, Lucas M, Seth NP et al. Transient immunological control during acute hepatitis C virus infection: *ex vivo* analysis of helper T-cell responses. J Viral Hepat. 2006; 13:708–14.
54. Scotto G, Fazio V, D'Alessandro G et al. Association between HLA class II antigens and hepatitis C virus infection. J Biol Regul Homeost Agents. 2003;17:316–21.
55. Yenigun A, Durupinar B. Decreased frequency of the HLA-DRB1*11 allele in patients with chronic hepatitis C virus infection. J Virol. 2002;76:1787–9.
56. Thursz M, Yallop R, Goldin R, Trepo C, Thomas HC. Influence of MHC class II genotype on outcome of infection with hepatitis C virus. The HENCORE group. Hepatitis C European Network for Cooperative Research. Lancet. 1999;354:2119–24.
57. Minton EJ, Smillie D, Neal KR, Irving WL, Underwood JC, James V. Association between MHC class II alleles and clearance of circulating hepatitis C virus. Members of the Trent Hepatitis C Virus Study Group. J Infect Dis. 1998;178:39–44.
58. Mangia A, Gentile R, Cascavilla I et al. HLA class II favors clearance of HCV infection and progression of the chronic liver damage. J Hepatol. 1999;30:984–9.
59. Zavaglia C, Martinetti M, Silini E et al. Association between HLA class II alleles and protection from or susceptibility to chronic hepatitis C. J Hepatol. 1998;28:1–7.
60. Alric L, Fort M, Izopet J et al. Genes of the major histocompatibility complex class II influence the outcome of hepatitis C virus infection. Gastroenterology. 1997;113:1675–81.
61. Wawrzynowicz-Syczewska M, Underhill JA, Clare MA, Boron-Kaczmarska A, McFarlane IG, Donaldson PT. HLA class II genotypes associated with chronic hepatitis C virus infection and response to alpha-interferon treatment in Poland. Liver. 2000;20:234–9.
62. Tibbs C, Donaldson P, Underhill J, Thomson L, Manabe K, Williams R. Evidence that the HLA DQA1*03 allele confers protection from chronic HCV-infection in Northern European Caucasoids. Hepatology. 1996;24:1342–5.

5
Liver cancer

C. ARMENGOL, S. CAIRO and M. A. BUENDIA

Primary liver tumours are classified into different histological types, depending on the lineage of transformed cells and clinicopathological parameters, as illustrated in Figure 1. Hepatocellular carcinoma (HCC) is the major type of liver cancer and one of the commonest human neoplasms worldwide[1]. HCC typically arises on a background of prolonged necroinflammatory liver disease, in frequent association with liver cirrhosis. It may be induced by multiple risk

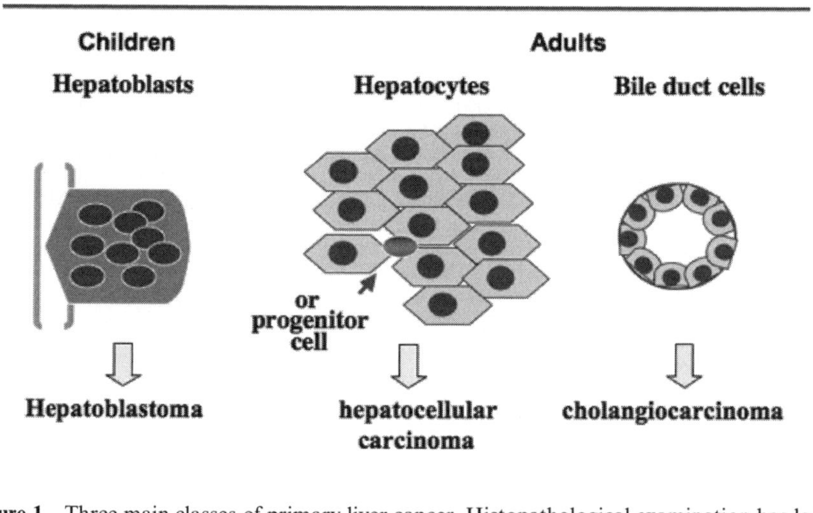

Figure 1 Three main classes of primary liver cancer. Histopathological examination has led to the identification of three classes of primary liver tumours that present similarities with different cell types in the normal liver, including cell morphology, tissue architecture and expression of specific protein markers. Hepatoblastoma is the predominant malignant hepatic neoplasm in young children and it appears to be derived from hepatoblasts, the bipotent precursor cells in fetal liver. Hepatocellular carcinoma (80–85% of primary liver cancers) results from malignant transformation of either mature hepatocytes, the major epithelial cell type in adult liver, or hepatic precursor cells similar to oval cells. Cholangiocarcinoma (about 10% of primary liver tumours) develops from the epithelium of the intrahepatic biliary tree

factors such as chronic viral infections and environmental carcinogens. In this context the interplay between various oncogenic pathways leading to liver cell transformation remains incompletely understood[2,3]. For HCC, as for other human cancers, the recent years have seen an explosion of data providing global insights into genetic alterations and gene expression profiles in large sets of tumours. Most of these studies, showing a large variety of genetic alterations and striking heterogeneity of gene expression profiles between individual HCC cases, have supported the notion that HCC ranks probably among the most complex and heterogeneous types of human solid tumours[4–6]. This idea also fits with the multiple aetiologies of HCC and the long period of chronic inflammatory disease that fosters the accumulation of genetic and epigenetic defects. As a consequence of the activation of cellular oncogenes or the inactivation of tumour-suppressor genes, deregulation of various signalling pathways has been reported in HCC subsets, such as Wnt/β-catenin signalling, the p14ARF/p53, p16INK4A/RB, TGF-β, and PTEN/Akt pathways (reviewed in ref. 7).

Aberrant reactivation of the developmental Wnt/β-catenin pathway is a predominant mechanism implicated in liver tumorigenesis[8]. The canonical Wnt/Wingless signalling pathway plays a pivotal role in regulating growth and cell fate in early and late stages of development[9]. As illustrated in Figure 2, these effects are achieved through the stabilization of β-catenin and its translocation to the nucleus as a coactivator for high-mobility group (HGM)-box proteins of the Tcf/Lef family. In the absence of Wnt, a multiprotein complex including the protein kinase GSK3β, adenomatous polyposis coli (APC) and Axin induces phosphorylation of β-catenin at N-terminal serine

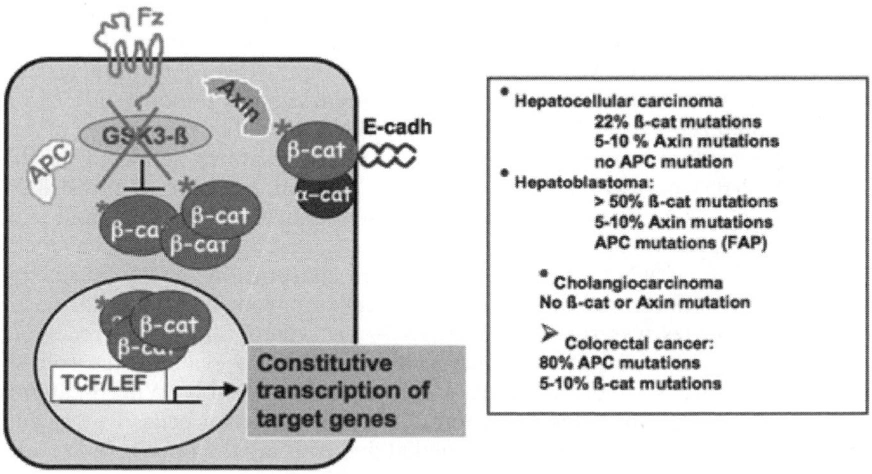

Figure 2 Activation of the Wnt/β-catenin pathway in cancer cells. Left panel: schematic illustration of the intracellular accumulation of mutated β-catenin leading to enhanced expression of Wnt target genes. Right panel: comparison of the rates of β-catenin activation in different types of primary liver cancer and in colorectal cancer[2]

and threonine residues, and phosphorylated β-catenin is directed towards proteasome-mediated degradation. Activation of Wnt abrogates the degradation pathway, leading to elevated levels of transcriptionally active β-catenin. Nuclear accumulation of β-catenin induces a transcriptional switch, in which Tcf-bound repressors (CtBP, TLE/Groucho, HDAC) are displaced by β-catenin and its associated coactivators CBP/p300, Brg-1, TIP49/Pontin-52 and Bcl9-pygopus[10]. The selective activation of distinct Wnt target genes in proper context is strictly controlled by the interplay of positive and negative regulatory signals on Wnt-responsive promoters.

Recent studies have outlined the importance of Wnt/β-catenin signalling in regulating liver cell proliferation during development and in governing essential functions in the adult liver[11–13]. Moreover, aberrant reactivation of Wnt signalling is a predominant mechanism implicated in liver tumorigenesis. As shown in Figure 2, mutations of the β-catenin and Axin genes leading to constitutive activation of β-catenin have been identified in HCC and hepatoblastoma (HB), but not in cholangiocarcinomas (CC)[14–16]. Animal models have been instrumental in allowing identification of β-catenin target genes in non-diseased or tumorous liver, including liver-specific enzymes involved in glutamine and nitrogen metabolism such as glutamine synthetase (GS), ornithine aminotransferase, the glutamate transporter (GLT-1), and the cytochrome P450 enzymes CYP1A2 and CYP2E1[13,17]. However, c-myc has not so far been found to be regulated by β-catenin in the liver context, and the oncogenic programme triggered by Wnt signalling in hepatocarcinogenesis remains largely unknown. We have recently found that the T-box protein Tbx3 is a direct transcriptional target of the β-catenin/Tcf4 complex and that it is specifically overexpressed in human and murine HCC harbouring a mutated β-catenin allele[18]. Tbx3 plays an important role in patterning events during development, and in humans, Tbx3 mutations are responsible for the ulnar–mammary syndrome (UMS), an autosomal dominant disorder characterized by upper limb deficiencies and apocrine/mammary gland hypoplasia[19]. As a transcriptional repressor of $p14^{ARF}$, Tbx3 has been linked to the control of cell cycle and senescence, and of the anti-proliferative response delivered by the ARF-Mdm2-p53 pathway in tumorigenesis[20]. Indeed, we have demonstrated that Tbx3 acts as a critical mediator of β-catenin survival functions and as a new, important player in liver cancer.

The question of whether HCC arises from maturation arrest caused by aberrant differentiation of determined stem cells or from dedifferentiation of mature hepatocytes was raised more than 10 years ago, and is still receiving considerable attention[6,21,22]. It is therefore important to evaluate the role of signalling pathways involved in liver development, notably Wnt/β-catenin signalling, in malignant transformation of hepatic precursor cells. To this end we have recently launched a project aimed at determining the impact of Wnt/β-catenin in the molecular pathogenesis of hepatoblastoma.

Hepatoblastoma is the most frequent primary liver tumour in children under 3 years of age, but it is a rare tumour representing less than 1% of all paediatric malignancies[23]. HB originate from immature liver precursor cells and present morphological features recapitulating some of the developmental aspects of the liver, with four patterns: fetal, embryonal, macrotrabecular, and small cell undifferentiated types[24]. Mixed epithelial–mesenchymal tumours also contain myofibroblastic, chondroid and osteoid tissues with occasional teratoid features, suggesting that they might derive from uncommitted stem cells or present intrinsic capacities to undergo epithelial–mesenchymal transition. Unlike HCC, hepatoblastomas develop in the absence of underlying liver disease and viral infection. Evidence for genetic/epigenetic origin of HB has been provided by its association with congenital disorders or overgrowth syndromes such as the Beckwith–Wiederman syndrome, with familial adenomatous polyposis (FAP), and with prematurity and low birth weight[25–27]. In sporadic cases genetic analysis has uncovered a high frequency of mutations in the β-catenin or Axin genes, leading to constitutive activation of the Wnt pathway in a majority of HB cases[16,28].

We have recently analysed a series of 70 hepatoblastomas collected from several French medical centres. Extensive pathological and clinical annotations were obtained, most patients being enrolled in the international clinical trials of the SIOPEL group[29]. The β-catenin mutational status was determined by RT-PCR over exons 3 and 4. Tumour proliferative index was determined by Ki67 labelling of fixed, paraffin-embedded tumour sections. Finally, we analysed gene expression profiles on Affymetrix chips and genetic alterations by array CGH in 30 tumours and four pools of normal livers from HB patients. Microarray data of significant genes were validated by Q-PCR and immunohistochemistry, and molecular signatures were validated in a second, independent set of 40 HB (Armengol, Cairo and Buendia, manuscript in preparation). Unsupervised hierarchical clustering clearly separated tumours from normal liver and identified two tumour groups: cluster A: localized tumours with fetal histology, low proliferative index and good response to chemotherapy; and cluster B: invasive tumours with less differentiated patterns, high proliferative index and unfavourable prognosis. The mutational status of β-catenin was not significantly associated with the two different tumour groups. However, we observed striking differences in subcellular localization of β-catenin and in deregulated β-catenin target genes between the two clusters (Figure 3). Overexpression of known β-catenin target genes related to fatty acid

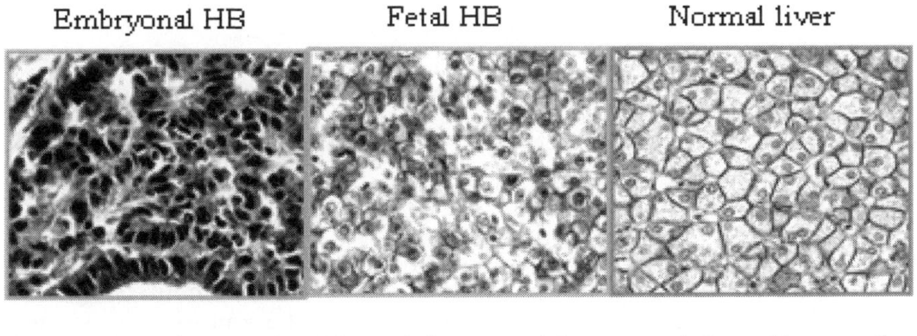

Figure 3 Illustrative examples of β-catenin immunostaining patterns in hepatoblastoma. Cases harbouring a mutated β-catenin are shown. Intracellular accumulation of the protein is seen both in embryonal and fetal tumours, with marked differences in membranous staining and intensity of nuclear staining. In normal hepatocytes, β-catenin is mainly localized at the plasma membrane

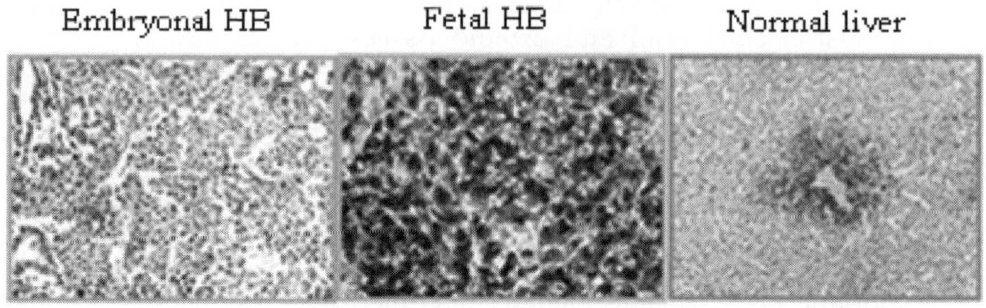

Figure 4 Immunostaining of glutamine synthetase (GS), a critical β-catenin target in the liver context, in different types of hepatoblastoma. Strong GS expression is commonly observed in fetal hepatoblastomas, but not in embryonal tumours. In the normal liver, GS expression is localized to hepatocytes in the perivenous area, corresponding to previously described metabolic zonation and the reported role of β-catenin in controlling the perivenous genetic programme[12]

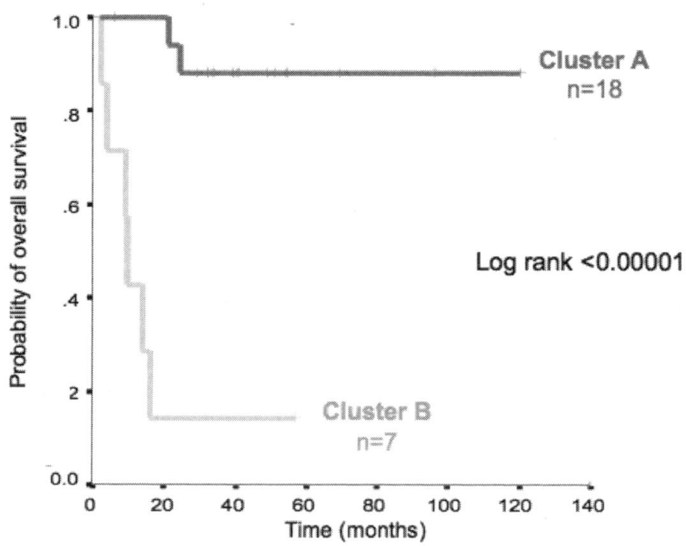

Figure 5 Correlation of molecular subtypes of hepatoblastoma with overall survival of patients. Kaplan–Meier curves show that cases in cluster B (poorly differentiated tumours) have reduced survival probability compared to cases in cluster A (well-differentiated, fetal-type tumours). Log-rank test, $p = 0.00001$

or amino acid metabolism, such as glutamine synthetase (GS), was predominant in cluster A, but not in less differentiated tumours (Figure 4). Conversely, β-catenin targets related to cell cycle and apoptosis were strongly activated in cluster B and down-regulated in cluster A.

Supervised analysis comparing clusters A and B revealed a significant enrichment of genes located on chromosomes 2 and 8q in cluster-B tumours. Moreover, in CGH array analysis, gains of chromosomes 2 and 8q were also significantly associated to cluster B. Importantly, distinct profiles of gene expression and genetic alterations in clusters A and B were significantly associated with several clinicopathological parameters. In particular, the 2-year probability of patient survival differed significantly between clusters A and B (90% vs 15%, $p<0.0001$). Furthermore, 18 differently expressed genes ($p<1\times10^{-7}$) between clusters A and B were identified as putative prognostic factors. Their predictive value was confirmed by QPCR in an independent series of 40 HB.

In conclusion, global analysis of gene expression profiles and genetic alterations in hepatoblastoma has revealed two distinct prognostic subtypes of tumours that are strongly related to tumour cell differentiation stage and proliferation, and to specific chromosomal alterations. It will be important to compare these data with those of human and murine HCC to improve our understanding of liver carcinogenesis. Most importantly, better understanding of the molecular pathogenesis of liver cancers has considerable implication for cancer prevention and therapy. Since abnormal activation of the Wnt/β-catenin pathway plays a fundamental role in HB pathogenesis, as well as in a subset of HCC, specific strategies aimed at inhibiting this pathway in tumour cells should provide interesting alternative approaches for tumours that are resistant to current chemotherapeutic treatments.

Acknowledgements

This project was supported by grants from the Ligue Nationale contre le Cancer (Carte d'Identité des Tumeurs Program) and from the GIS Institut des Maladies Rares, and by fellowships from the Fondation pour la Recherche Médicale (C.A. and S.C.).

References

1. Bruix J, Boix L, Sala M, Llovet JM. Focus on hepatocellular carcinoma. Cancer Cell. 2004; 5:215–19.
2. Buendia MA. Genetics of hepatocellular carcinoma. Semin Cancer Biol. 2000;10:185–200.
3. Thorgeirsson SS, Grisham JW. Molecular pathogenesis of human hepatocellular carcinoma. Nat Genet. 2002;31:339–46.
4. Laurent-Puig P, Legoix P, Bluteau O et al. Genetic alterations associated with hepatocellular carcinomas define distinct pathways of hepatocarcinogenesis. Gastroenterology. 2001;120:1763–73.
5. Lee JS, Chu IS, Heo J et al. Classification and prediction of survival in hepatocellular carcinoma by gene expression profiling. Hepatology. 2004;40:667–76.
6. Lee JS, Heo J, Libbrecht L et al. A novel prognostic subtype of human hepatocellular carcinoma derived from hepatic progenitor cells. Nat Med. 2006;12:410–16.
7. Lee JS, Thorgeirsson SS. Comparative and integrative functional genomics of HCC. Oncogene. 2006;25:3801–9.
8. Giles RH, van Es JH, Clevers H. Caught up in a Wnt storm: Wnt signaling in cancer. Biochim Biophys Acta. 2003;1653:1–24.
9. Cadigan KM, Nusse R. Wnt signaling: a common theme in animal development. Genes Dev. 1997;11:3286–305.
10. Thompson B, Townsley F, Rosin-Arbesfeld R, Musisi H, Bienz M. A new nuclear component of the Wnt signalling pathway. Nat Cell Biol. 2002;4:367–73.

11. Monga SP, Monga HK, Tan X, Mule K, Pediaditakis P, Michalopoulos GK. Beta-catenin antisense studies in embryonic liver cultures: role in proliferation, apoptosis, and lineage specification. Gastroenterology. 2003;124:202–16.
12. Benhamouche S, Decaens T, Godard C et al. Apc tumor suppressor gene is the 'zonation-keeper' of mouse liver. Dev Cell. 2006;10:759–70.
13. Sekine S, Lan BY, Bedolli M, Feng S, Hebrok M. Liver-specific loss of beta-catenin blocks glutamine synthesis pathway activity and cytochrome p450 expression in mice. Hepatology. 2006;43:817–25.
14. de La Coste A, Romagnolo B, Billuart P et al. Somatic mutations of the beta-catenin gene are frequent in mouse and human hepatocellular carcinomas. Proc Natl Acad Sci USA. 1998;95:8847–51.
15. Tran van Nhieu J, Renard CA, Wei Y, Cherqui D, Zafrani ES, Buendia MA. Nuclear accumulation of mutated beta-catenin in hepatocellular carcinoma is associated with increased cell proliferation [In process citation]. Am J Pathol. 1999;155:703–10.
16. Wei Y, Fabre M, Branchereau S, Gauthier F, Perilongo G, Buendia MA. Activation of beta-catenin in epithelial and mesenchymal hepatoblastomas. Oncogene. 2000;19:498–504.
17. Cadoret A, Ovejero C, Terris B et al. New targets of beta-catenin signaling in the liver are involved in the glutamine metabolism. Oncogene. 2002;21:8293–301.
18. Renard CA, Labalette C, Armengol C et al. Tbx3 Is a downstream target of the Wnt/beta-catenin pathway and a critical mediator of beta-catenin survival functions in liver cancer. Cancer Res. 2007;67:901–10.
19. Bamshad M, Lin RC, Law DJ et al. Mutations in human TBX3 alter limb, apocrine and genital development in ulnar-mammary syndrome. Nat Genet. 1997;16:311–15.
20. Brummelkamp TR, Kortlever RM, Lingbeek M et al. TBX-3, the gene mutated in ulnar-mammary syndrome, is a negative regulator of p19ARF and inhibits senescence. J Biol Chem. 2002;277:6567–72.
21. Sell S. Cellular origin of cancer: dedifferentiation or stem cell maturation arrest? Environ Health Perspect. 1993;101(Suppl. 5):15–26.
22. Roskams T. Liver stem cells and their implication in hepatocellular and cholangiocarcinoma. Oncogene. 2006;25:3818–22.
23. Schnater JM, Kohler SE, Lamers WH, von Schweinitz D, Aronson DC. Where do we stand with hepatoblastoma? A review. Cancer. 2003;98:668–78.
24. Zimmermann A. The emerging family of hepatoblastoma tumours: from ontogenesis to oncogenesis. Eur J Cancer. 2005;41:1503–14.
25. DeBaun MR, Tucker MA. Risk of cancer during the first four years of life in children from the Beckwith–Wiedemann Syndrome Registry. J Pediatr. 1998;132:398–400.
26. Tanimura M, Matsui I, Abe J et al. Increased risk of hepatoblastoma among immature children with a lower birth weight. Cancer Res. 1998;58:3032–5.
27. Ansell P, Mitchell CD, Roman E, Simpson J, Birch JM, Eden TO. Relationships between perinatal and maternal characteristics and hepatoblastoma: a report from the UKCCS. Eur J Cancer. 2005;41:741–8.
28. Koch A, Denkhaus D, Albrecht S, Leuschner I, von Schweinitz D, Pietsch T. Childhood hepatoblastomas frequently carry a mutated degradation targeting box of the b-catenin gene. Cancer Res. 1999;59:269–73.
29. Perilongo G, Shafford E, Maibach R et al. Risk-adapted treatment for childhood hepatoblastoma. final report of the second study of the International Society of Paediatric Oncology – SIOPEL 2. Eur J Cancer. 2004;40:411–21.

6
Hans Popper Aware Lecture: Circulatory and renal failure in cirrhosis

A. CÁRDENAS and P. GINÈS

INTRODUCTION

Patients with advanced cirrhosis and portal hypertension often display a profound disturbance in the systemic circulation, characterized by a low arterial blood pressure, high cardiac output and a decreased total systemic vascular resistance[1,2]. This haemodynamic pattern is a result of an intense arterial vasodilation occurring in the splanchnic vascular bed that triggers a neurohormonal response causing haemodynamic changes characteristic of cirrhosis. By contrast, other vascular beds such as the cerebral, upper and lower limbs and renal circulation, are constricted in patients with advanced cirrhosis[2–4]. These circulatory abnormalities increase with the progression of liver disease, and become very pronounced in patients with advanced cirrhosis. One of the main consequences of these changes is the development of functional renal failure which is also known as vasodilation-induced renal failure. This type of renal failure includes hepatorenal syndrome (HRS) and renal dysfunction related to sepsis and bacterial infections in cirrhosis. The main characteristic of HRS is the presence of renal vasoconstriction; which occurs late in the natural history of patients with cirrhosis and ascites and is a consequence of a continuous process in which several underlying mechanisms, including changes in systemic arterial circulation, increased portal pressure, impaired cardiac function and activation of systemic and renal vasoconstrictor factors, act on the renal circulation and lead to functional renal failure without histological damage of the kidneys. This chapter will describe the clinical aspects of circulatory failure in patients with cirrhosis. In addition a review of the different types of renal failure present in cirrhosis and their management will be discussed.

CIRCULATORY FUNCTION IN CIRRHOSIS

Arterial vasodilation

Portal hypertension due to an abnormal liver architecture and increased intrahepatic resistance to blood flow is the main factor responsible for the development of splanchnic arterial vasodilation. This vasodilation occurs mainly due to the production of nitric oxide and other vasodilators (calcitonin gene-related peptide, substance P, carbon monoxide, and endogenous cannibinoids, among others) as a consequence of endothelial stretching and possibly bacterial translocation[5]. In the early stages of cirrhosis (compensated state), splanchnic arterial vasodilation leads to a decreased effective arterial blood volume that is compensated for by an increased plasma volume and cardiac output that maintain a normal arterial pressure[6] (Figure 1).

As the disease progresses, plasma volume accumulates in the splanchnic bed, causing a compensatory response due to a fall in central blood volume with activation of systemic vasoconstrictor and antinatriuretic systems such as the renin–angiotensin–aldosterone system (RAAS), the sympathetic nervous system (SNS) and arginine vasopressin (AVP) accounting for sodium and water retention and in very advanced stages; renal vasoconstriction as the kidney senses a relative hypovolaemic state (Figure 2)[7–10]. The ongoing action of such vasoconstrictor factors acting on the kidney eventually leads to HRS. This occurs because, although in the early stages of cirrhosis renal blood flow

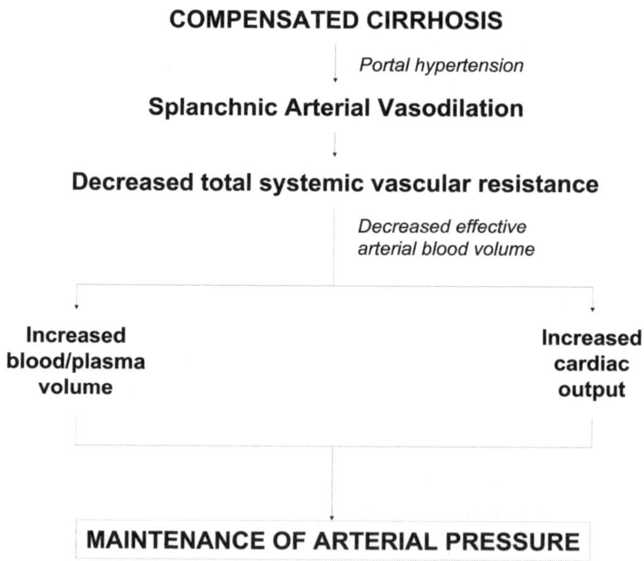

Figure 1 Circulatory changes in compensated cirrhosis. Splanchnic arterial vasodilation leads to a decreased effective arterial blood volume that is compensated for by an increased plasma volume and cardiac output that maintain a normal arterial blood pressure

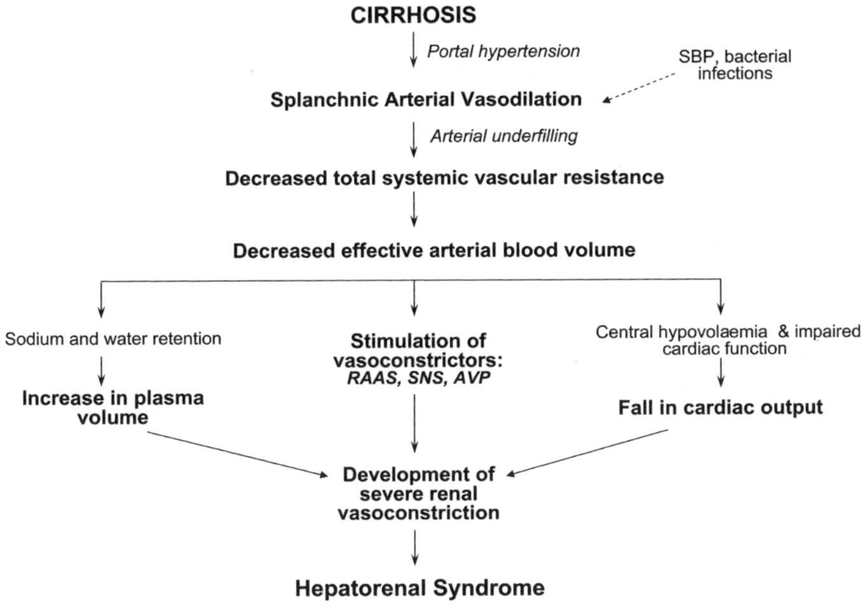

Figure 2 The pathogenesis of vasodilation-induced renal failure in patients with decompensated cirrhosis. The main precipitating factors are spontaneous bacterial peritonitis (SBP), and other bacterial infections. Splanchnic vasodilation arising from portal hypertension, an increased plasma volume and a decreased cardiac output seems to play an equally important role in the decreased renal perfusion leading to HRS. Renin–angiotensin–aldosterone system (RAAS), sympathetic nervous system (SNS), and arginine vasopressin (AVP)

may be kept within normal limits due to the effect of local renal vasodilators, with time circulating vasoconstrictors overcome the effect of renal vasodilators leading to severe renal vasoconstriction and reduction in glomerular filtration rate (GFR)[11]. In some patients precipitating factors such as bacterial infections worsen circulatory dysfunction and may trigger renal vasoconstriction[12–14] (Figure 2). At this stage the stimulation of the RAAS, SNS and AVP is so intense that their vasoconstrictor effect cannot be overcome and HRS develops.

Cardiac output

Circulatory dysfunction in cirrhosis has been the classic defining pathophysiological alteration that leads to HRS; however, recent evidence suggests that a reduction in cardiac output may also play a role[15,16]. As discussed above, a complex interplay between the splanchnic, systemic and renal circulation takes place once portal hypertension is established. In the initial stages of cirrhosis in the process of progressive vasodilation, both intravascular volume and cardiac output increase to maintain systemic haemodynamic homeostasis. However, with progression to decompensated cirrhosis a patient's clinical status may worsen, for instance due to a bacterial infection, and a subsequent decrease in cardiac output and effective arterial blood volume may occur, thereby reducing renal perfusion and possibly contributing to the development of HRS. Two studies suggest that the development of HRS occurs in the setting of a reduction in cardiac output, indicating that the progression of circulatory and renal dysfunction in cirrhosis is not only due to splanchnic vasodilation but also to a fall in cardiac output[15,16]. In the first study, consisting of 23 patients with spontaneous bacterial peritonitis (SBP), eight developed renal failure and had a significantly reduced cardiac output compared to those who did not develop renal failure[15]. In the second study, consisting of 66 patients with cirrhosis and ascites that were longitudinally followed, 27 (41%) developed HRS and in all cases cardiac output at the time of HRS was significantly decreased in the same patients compared to the cardiac output before the occurrence of HRS[16]. This supports the concept that HRS occurs in the setting of worsening circulatory function with a significant decrease in cardiac output in the setting of marked splanchnic vasodilation[15,16]. The exact pathogenic mechanisms that cause this decrease in cardiac output are not completely understood and therefore deserve further investigation.

RENAL FAILURE IN CIRRHOSIS

Definition

The diagnosis of renal failure in patients with cirrhosis has been arbitrarily defined when serum creatinine increases over 1.5 mg/dl (133 μmol/L)[17]. In patients with cirrhosis a serum creatinine concentration of 1.5 mg/dl corresponds to a GFR of approximately 30 ml/min[17]. Although there are clearance techniques (i.e. inulin clearance) with greater sensitivity than that of serum creatinine concentration for the measurement of GFR, they are cumbersome, expensive and not available in all settings. For these reasons serum creatinine concentration is currently considered the method of choice to estimate GFR in cirrhosis in clinical practice[17]. Although a single measurement of serum creatinine concentration may underestimate GFR, because of the existence of a decreased endogenous production of creatinine in cirrhosis due to reduced muscle mass[18], repeated measurements of serum creatinine concentration over time are useful in indicating variations of GFR in clinical practice.

Causes of renal failure in cirrhosis (Table 1)

The most common cause of renal failure in cirrhosis is vasodilation-induced renal failure. This type of renal failure encompasses HRS and sepsis-related renal failure with bacterial infections. Other common causes of renal failure in cirrhosis include prerenal renal failure secondary to volume depletion, acute tubular necrosis (ATN) secondary to shock, drug-induced renal failure mainly from non-steroidal anti-inflammatory drugs (NSAID) or aminoglycosides and glomerulopathies in patients with cirrhosis due to hepatitis B or C or alcohol.

Table 1 Main causes of renal failure in patients with cirrhosis

1. Vasodilation-induced renal failure
   (a) Hepatorenal syndrome
   (b) Bacterial infections
2. Hypovolaemia-induced renal failure
   (a) Vomiting, diarrhoea
   (b) Gastrointestinal bleeding
   (c) Diuretic
3. Drug-induced renal failure
   (a) Non-steroidal anti-inflammatory drugs
   (b) Aminoglycosides
4. Intrinsic renal disease
   (a) Glomerulopathies due to viral hepatitis or alcohol

Vasodilation-induced renal failure

The pathogenesis of HRS and sepsis-related renal failure is the ultimate consequence of an unrelenting activation of vasoconstrictor factors acting on the kidney secondary to splanchnic vasodilation. Although in the early stages of cirrhosis renal blood flow may be kept within normal limits, due to the effect of renal vasodilators, with time circulating vasoconstrictors overcome the effect of renal vasodilators, leading to severe renal vasoconstriction and reduction in GFR[7,11]. In some patients precipitating factors such as bacterial infections worsen circulatory dysfunction and aggravate renal vasoconstriction. It is estimated that bacterial infections induce renal failure in 30–33% of patients with advanced cirrhosis[12–14]. The prevalence of renal failure seems to be higher in those patients who develop SBP, gastrointestinal and/or biliary infections[12–14]. Once this occurs, intrarenal mechanisms probably perpetuate HRS due to the development of intrarenal vicious cycles in which hypoperfusion leads to an imbalance in intrarenal vasoactive systems that in turn cause more vasoconstriction. It is important to emphasize that in this type of renal failure there is marked dilation of the arterial tree without reduction in intravascular blood volume, which in fact is increased[7]. The activation of vasoconstrictor systems also results in vasoconstriction of some vascular beds other than the kidneys, including the arms, legs, and brain[2–4]. The splanchnic area escapes from the effect of vasconstrictors, probably because of an increased local production of vasodilators.

Another important concept to consider in cirrhotic patients with circulatory failure is the development of adrenal insufficiency in the setting of advanced liver disease. Patients with cirrhosis and sepsis-related infections including SBP develop adrenal insufficiency in 50–70% of cases[19,20]. The development of adrenal insufficiency is associated with haemodynamic instability and increased mortality (80% in those with adrenal insufficiency vs 37% without adrenal insufficiency)[19]. Furthermore, in cirrhotic patients with adrenal insufficiency (diagnosed by the short corticotropin test within the first 24 h of admission) and septic shock, administration of intravenous hydrocortisone (50 mg every 6 h) helps with the resolution of shock and improves survival in patients with Child C cirrhosis[20].

Hypovolaemia-induced renal failure

Another type of renal failure that may occur in cirrhosis is pre-renal renal failure. This type of renal dysfunction is related to a reduction in intravascular volume leading to a fall in renal plasma flow and GFR[21]. Patients with cirrhosis are frequently exposed to clinical conditions that may cause a reduction in intravascular volume. These conditions include gastrointestinal bleeding, overdiuresis due to excessive diuretic treatment, and gastrointestinal fluid losses due to vomiting and/or diarrhoea. If renal failure is secondary to volume depletion, renal function may improve after volume repletion and treatment of the precipitating factor. In contrast to vasodilation-induced renal failure, in which there is arterial vasodilation without reduction in intravascular volume, in hypovolaemia-induced renal failure the major mechanism responsible for renal hypoperfusion is a reduction in intravascular volume.

Intrinsic renal diseases

Some intrinsic renal diseases may be present in patients with cirrhosis. Most of these are related to common aetiological factors of cirrhosis, including chronic hepatitis B or C infection or alcoholic liver disease, and encompass kidney diseases secondary to the deposition of circulating immunocomplexes in the glomeruli[22,23]. In some patients the glomerular deposits are mild and do not affect kidney function, whereas in other patients they are so severe that glomerular function is disrupted, GFR decreases and renal failure develops.

Drug-induced renal failure

Patients with cirrhosis have an increased risk of development of renal failure when treated with drugs that alter the equilibrium between vasodilator and vasoconstrictor factors in the renal circulation. Cirrhotic patients may develop renal failure when treated with a variety of drugs, especially NSAID, aminoglycosides, diuretics or vasodilators[24–26]. Of these, the most important and commonly used in cirrhotics are NSAID and diuretics.

NSAID inhibit the enzymes cycloxygenase-1 and 2 which are responsible for prostaglandin synthesis. Prostaglandins are important renal vasodilators that contribute significantly towards maintaining normal renal perfusion. The risk of developing renal failure due to NSAID administration is higher in patients with cirrhosis and ascites and increased activity of the vasoconstrictor systems, in which renal perfusion is dependent on an adequate renal synthesis of vasodilator prostaglandins, compared to that of patients with ascites without increased activity of vasoconstrictor systems[26]. Renal failure has been reported to occur after a short treatment with a variety of NSAID, including indomethacin, aspirin, ibuprofen, naproxen, and sulindac[27–29]. In these studies the GFR quickly returned to pretreatment values after cessation of the drug. However, it is not known whether renal failure may be reversible in patients treated for longer periods. Studies in patients with cirrhosis and ascites indicate that cycloxygenase-2 enzyme (COX-2) inhibitors do not cause renal failure when administered for a short period of time[29,30].

The incidence of renal impairment during diuretic treatment ranges between 20% and 40%[26,31]. This diuretic-induced renal failure is usually moderate and reversible after diuretic withdrawal, and is related to an imbalance between the fluid loss from the intravascular space caused by diuretic treatment and the passage of fluid from the peritoneal compartment to the general circulation. Patients with cirrhosis are also at high risk of developing renal failure due to ATN when treated with aminoglycosides[24,32]. Because of this high risk of nephrotoxicity and the existence of other effective antibiotics (i.e third-generation cephalosporins) treatment with aminoglycosides should be avoided in patients with chronic liver disease.

DIFFERENTIAL DIAGNOSIS OF RENAL FAILURE

There are no specific tests that secure the diagnosis of the different types of renal failure in cirrhosis. An algorithm for the differential diagnosis of renal failure in cirrhosis is shown in Figure 3. According to this algorithm the differential diagnosis of renal failure in patients with cirrhosis includes four steps.

1. *Investigation of treatment with nephrotoxic drugs.* The first step consists of the investigation of treatment with potentially nephrotoxic drugs; particularly NSAID, vasodilators, or aminoglycosides. The frequency and risk factors of renal failure in patients receiving these drugs is poorly known, but it appears that patients with severe circulatory dysfunction and severely impaired liver function are those with the highest risk of renal failure when treated with these drugs.

2. *Investigation of volume losses.* The second step consists of the clinical assessment of significant volume losses that could have led to hypovolaemia, including gastrointestinal bleeding, excessive diuresis, and loss of gastrointestinal fluid, due to vomiting or diarrhoea. Data from patients with cirrhosis and gastrointestinal bleeding indicate that an important reduction in blood volume (in most cases associated to hypovolaemic shock) is required to induce renal failure[33]. Therefore it is unlikely that small reductions in blood volume could lead to renal failure. This suggests that specific measures to assess blood volume status, such as measurement of central venous pressure, are not required in all patients with renal failure unless there is significant history of volume losses.

3. *Investigation of bacterial infections.* Bacterial infections are very common in patients with advanced cirrhosis and represent the most frequent cause of renal failure[12–14]. From a clinical standpoint the presence of a bacterial infection should be sought after in all patients with cirrhosis presenting with

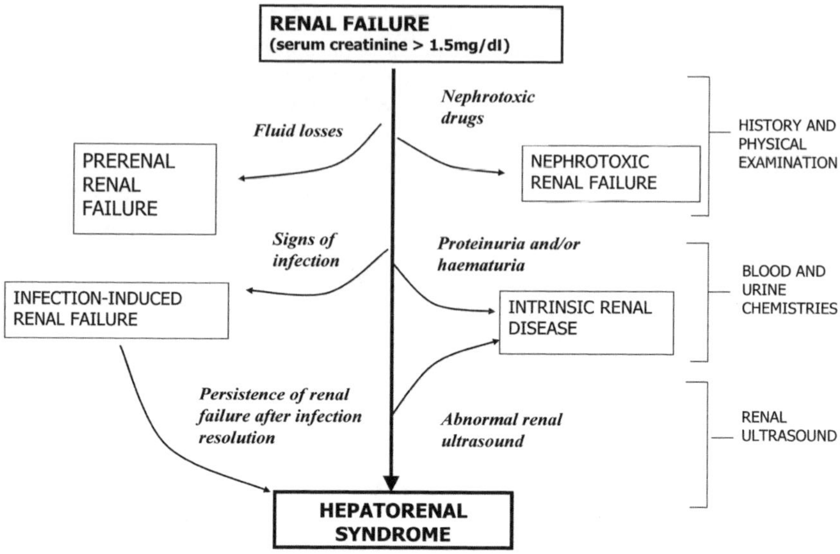

Figure 3 Diagnostic algorithm of renal failure in cirrhosis (Modified from Ginès et al. Lancet. 2003;362:1819-27)

renal failure. SBP is probably the most common cause, but other types of infection should also be investigated. It is important to mention that signs and symptoms of early bacterial infection may be vague or even absent in some patients with cirrhosis. Moreover, it is not uncommon that in hospitalized patients renal failure is detected before symptoms of infection become clinically evident. Therefore, objective signs of infection should be sought in all patients with cirrhosis and renal failure. These include blood cell count, with measurement of band forms, polymorphonuclear cell count in a sample of ascitic fluid, urine sediment, chest X-ray and cultures of blood and other organic fluids.

4. *Investigation of intrinsic renal diseases.* Finally, in patients with cirrhosis and renal failure due to intrinsic renal diseases, there is usually proteinuria or haematuria. Because a renal biopsy cannot be performed in all patients with cirrhosis and renal failure, to confirm the existence of glomerular diseases, significant proteinuria or haematuria is usually used as a surrogate marker of the presence of a glomerular disease.

If after extensive investigation there is no evidence of administration of nephrotoxic drugs, volume losses, bacterial infection or intrinsic renal diseases, then HRS should be considered the cause of renal failure (Figure 3).

MANAGEMENT

General measures

Moderate reductions of sodium and fluid intake are probably beneficial in all patients with cirrhosis and renal failure, except in those with renal failure due to hypovolaemia (see later), because most patients have sodium retention with ascites and oedema and dilutional hyponatraemia and the presence of renal failure causes a further impairment in renal sodium and solute-free water excretion.

Cardiopulmonary function should be assessed by clinical examination, although patients with cirrhosis and renal failure usually do not develop signs of volume overload unless treated aggressively with plasma expanders. Serum electrolyte levels should be measured frequently in all patients with renal failure. Hyperkalaemia is common but usually moderate. Potassium-sparing diuretics should not be given to these patients, to avoid marked increases in serum potassium levels. Metabolic acidosis is common but its severity depends on the intensity and cause of renal failure. Metabolic acidosis is usually severe in patients with renal failure due to shock, whereas it is usually moderate in most patients with HRS.

Patients with cirrhosis and renal failure are at high risk for other complications of cirrhosis. Hepatic encephalopathy is common in the setting of renal failure and patients should be managed with standard measures to treat this complication. SBP and other bacterial infections are very common in patients with advanced cirrhosis and represent a major cause of death; therefore cultures and a low threshold for suspecting infections should be performed. Finally, malnutrition is also common and attempts should be made to increase calorie intake.

Specific measures

Renal failure due to nephrotoxic drugs

Renal failure due to the administration of NSAID is reversible upon drug withdrawal. Nevertheless, it should be emphasized that the existing information is limited to pathophysiological studies in which NSAID were given during a short period of time (usually less than 7 days)[27–30]. Therefore, the outcome of patients who have been treated with NSAID for longer periods of time is not known. The information on nephrotoxicity due to vasodilators and aminoglycosides is also limited, and no specific effective treatment measures other than drug withdrawal have been reported.

Renal failure due to hypovolaemia

Management of patients with hypovolaemia-induced renal failure depends on the cause of volume depletion. In patients with gastrointestinal bleeding, early resuscitation with plasma expanders and packed red blood cells, together with effective measures to stop bleeding and maintain arterial pressure, are essential to prevent the development of renal failure.

In patients with ascites the cautious use of diuretics is fundamental to the prevention of renal failure due to excessive diuresis. This is particularly important in patients using them for the first time; these individuals usually have an excellent response to low doses of diuretics. When starting a diuretic regimen in a patient with cirrhosis, visits should be scheduled frequently (once or twice a week) during the initiation of treatment and the patient instructed to reduce the dose of diuretics if weight loss is greater than expected (0.5 kg/day in patients without oedema and 1 kg/day in patients with oedema). Diuretic doses should be reduced before the complete disappearance of ascites and oedema to prevent a decrease in plasma volume and subsequent renal failure. In patients presenting with renal failure due to excessive diuresis, diuretics should be stopped immediately. Specific measures in these patients include the administration of isotonic saline, until signs of volume depletion disappear.

Renal failure due to bacterial infections

In patients with SBP, administration of albumin (1.5 g/kg body weight at the diagnosis of the infection and 1 g/kg body weight at day 3) together with the antibiotic therapy reduces the incidence of renal failure from approximately 30% in patients not receiving albumin to 10%[13]. This beneficial effect of albumin on renal function is associated with an improved survival. In patients with SBP who develop renal failure despite the administration of albumin, renal failure should be treated with vasoconstrictors and albumin (see treatment of HRS). Although it is possible that albumin administration is effective in reducing the incidence of renal failure in patients with cirrhosis and sepsis unrelated to SBP, no studies have so far been reported assessing this possibility.

Renal failure due to intrinsic renal diseases

There is no specific therapy for patients with cirrhosis and renal failure due to intrinsic kidney diseases such as glomerulonephritis, interstitial nephritis, diabetic nephropathy or hypertensive nephropathy. Drugs inhibiting the activity of the renin–angiotensin system should not be used in patients with cirrhosis because these drugs usually induce arterial hypotension and may impair renal function further. Antiviral therapy may be effective in some cases of glomerulonephritis associated with hepatitis B or C infection[22]. However, antiviral drugs should be used with caution because most of these patients have advanced liver disease, which may increase the incidence and severity of side-effects related to these drugs. In patients with severe renal failure, haemodyalisis or other forms of renal replacement therapy may be used. The beneficial effects of haemodialysis in these patients should be weighed against possible adverse effects, including hypotension, bacterial infections, and gastrointestinal bleeding. As a general rule, haemodialysis is indicated in suitable candidates for liver transplantation who can benefit from combined liver–kidney transplantation. In patients not candidates for liver transplantation, the decision of performing haemodialysis should be individualized in each patient.

Hepatorenal syndrome

HRS can occur in two clinical patterns: a rapidly progressive form, also known as type 1 HRS, and a stable or slowly progressive form, known as type 2 HRS[17]. Most of the information on therapy of HRS refers to type 1 HRS, while there is less information on therapy of type 2 HRS. The most important aspect of providing care to patients with HRS is assessment of candidacy for orthotopic liver transplantation. In those who are candidates for transplantation, all efforts should be made to improve renal function in order to obtain a better outcome after transplantation

Vasoconstrictor drugs (terlipressin or alpha-adrenergic agents) in combination with albumin to counteract splanchnic arterial vasodilation may improve circulatory and renal function in 50–75% of patients[34–44]. Octreotide is ineffective when administered alone[45], yet it has been reported to be beneficial when given in combination with albumin and midodrine[46–48]. Whether octreotide improves the efficacy of midodrine is unknown. Recurrence of HRS is uncommon after discontinuation of vasoconstrictors, although it is unknown whether the recurrence rate differs between patients with type 1 HRS and those with type 2. Patients who have a response to terlipressin have a higher survival rate than patients who do not have a response[36,37,47]; therefore, treatment with vasoconstrictors may increase the likelihood that patients with HRS will survive long enough to undergo liver transplantation. In addition, these drugs offer the advantage of improving renal function before transplantation, a benefit that may reduce post-transplant morbidity and mortality rates[49]. Although emerging data on the use of vasoconstrictors in patients with HRS are very promising, the available information is still limited and is based only on non-randomized studies involving small numbers of patients and with short follow-up. A list of recommendations regarding the use of vasoconstrictors in HRS is described in Table 2.

Table 2 Recommendations for using vasoconstrictors in patients with type 1 hepatorenal syndrome

Goal
Reduction of serum creatinine concentrations to $\leqslant 1.5$ mg/dl.
Recommendations
Terlipressin 0.5–1 mg should be given intravenously every 4 h; the dose can be increased in a stepwise fashion (i.e. every 2 days) to 1–2 mg/4 h in cases where there is no reduction in serum creatinine concentration.
Midodrine 2.5–7.5 mg should be given orally three times daily, with an increase to 12.5 mg three times daily if needed;octreotide 100 μg should be given subcutaneously three times daily, with an increase to 200 μg three times daily if needed.
Norepinephrine titrated to 0.5–3 mg/h as a continuous intravenous infusion.
Concomitant intravenous albumin infusion (1 g/kg on the first day, followed by 20–50 g/day) is recommended.
Contraindications include coronary artery disease, peripheral vascular disease, and/or cerebrovascular disease, due to the potential risk of ischaemic events.
The duration of therapy should be aproximately 7–14 days.

Transjugular intrahepatic portosystemic shunting (TIPS), a non-surgical intervention performed percutaneously that creates a connection (by means of a prosthetic shunt) within the liver between the portal and hepatic vein, thereby reducing portal pressure, also appears to be effective in treating HRS, yet the existing information is insufficient. There are scarce data regarding the use of TIPS in HRS. Four uncontrolled studies indicate that TIPS may improve renal function and GFR as well as reducing the activity of RAAS and SNS in patients with cirrhosis and HRS[47,50–52]. Haemodialysis should not be used routinely in patients with HRS because it does not improve outcome. However, it may have a role as a bridge to liver transplantation in patients who do not respond to medical therapy.

References

1. Cárdenas A, Sánchez-Fueyo A. Circulatory dysfunction in cirrhosis. Physiopathology and clinical implications. Gastroenterol Hepatol. 2003;26:447–55.
2. Menon K, Kamath P. Regional and systemic hemodynamic disturbances in cirrhosis. Clin Liver Dis. 2001;5:617–27.
3. Guevara M, Bru C, Ginès P et al. Increased cerebral vascular resistance in cirrhotic patients with ascites. Hepatology. 1998;28:39–44.
4. Maroto A, Ginès P, Arroyo V et al. Brachial and femoral artery blood flow in cirrhosis: relationship to kidney dysfunction. Hepatology. 1993;17:788–93.
5. Iwakiri Y, Groszmann R. The hyperdynamic circulation of chronic liver diseases: from the patient to the molecule. Hepatology. 2006;43:S121–31.
6. Cárdenas A, Ortega R, Ginès P. The hepatocirculatory syndrome and the management of patients with ascites. In: Arroyo V, Ginès P, Navasa M, Bruix J, Bosch J, Rodes J, editors. Therapy in Hepatology. Barcelona: Ars Medica, 2001:33–42.
7. Schrier RW, Arroyo V, Bernardi M et al. Peripheral arterial vasodilation hypothesis: a proposal for the initiation of renal sodium and water retention in cirrhosis. Hepatology. 1988;8:1151–7.
8. Bernardi M, Domenicali M. The renin–angiotensin–aldosterone system in cirrhosis. In: Ginès P, Arroyo V, Rodés J, Schrier RW, editors. Ascites and Renal Dysfunction in Liver Disease. Oxford: Blackwell, 2005:43–53.
9. Dudley F, Esler M. The sympathetic nervous system in cirrhosis. In: Ginès P, Arroyo V, Rodés J, Schrier RW, editors. Ascites and Renal Dysfunction in Liver Disease. Oxford: Blackwell, 2005:54–72.
10. Schrier RW. Water and sodium retention in edematous disorders: role of vasopressin and aldosterone. Am J Med. 2006;119:S47–53.
11. Arroyo V, Terra C, Gines P. New treatments of hepatorenal syndrome. Semin Liver Dis. 2006;26:254–64.
12. Terra C, Guevara M, Torre A et al. Renal failure in patients with cirrhosis and sepsis unrelated to spontaneous bacterial peritonitis. Value of MELD score. Gastroenterology. 2005;129:1944–53.
13. Sort P, Navasa M, Arroyo V et al. Effect of intravenous albumin on renal impairment and mortality in patients with cirrhosis and spontaneous bacterial peritonitis. N Engl J Med. 1999;5:403–9.
14. Fasolato S, Angeli P, Dallagnese L et al. Renal failure and bacterial infections in patients with cirrhosis: epidemiology and clinical features. Hepatology. 2007;45:223–9
15. Ruiz-del-Arbol L, Urman J, Fernandez J et al. Systemic, renal, and hepatic hemodynamic derangement in cirrhotic patients with spontaneous bacterial peritonitis. Hepatology. 2003; 38:1210–18.
16. Ruiz del Arbol L, Monescillo A, Arocena C et al. Circulatory function and hepatorenal syndrome in cirrhosis. Hepatology. 2005;42:439–47.
17. Arroyo V, Ginès P, Gerbes AL et al. Definition and diagnostic criteria of refractory ascites and hepatorenal syndrome in cirrhosis. Hepatology. 1996;23:164–76.

18. Caregaro L, Menon F, Angeli P et al. Limitations of serum creatinine level and creatinine clearance as filtration markers in cirrhosis. Arch Intern Med. 1994;154:201–5.
19. Tsai MH, Peng YS, Chen YC et al. Adrenal insufficiency in patients with cirrhosis, severe sepsis and septic shock. Hepatology. 2006;43:673–81.
20. Fernández J, Escorsell A, Zabalza M et al. Adrenal insufficiency in patients with cirrhosis and septic shock: effect of treatment with hydrocortisone on survival. Hepatology. 2006: 44:1288–95.
21. Lameire N, Van Biesen W, Vanholder R. Acute renal failure. Lancet. 2005;365:417–30.
22. Poole B, Schrier RW, Jani A. Glomerular disease in cirrhosis. In: Ginès P, Arroyo V, Rodés J, Schrier, editors. Ascites and Renal Dysfunction in Liver Diseases. Oxford: Blackwell, 2005:360–71.
23. Meyers CM, Seeff LB, Stehman-Breen CO, Hoofnagle JH. Hepatitis C and renal disease: an update. Am J Kidney Dis. 2003;42:631–57.
24. Cabrera J, Arroyo V, Ballesta AM et al. Aminoglycoside nephrotoxicity in cirrhosis. Value of urinary beta-2 microglobulin to discriminate functional renal failure from acute tubular damage. Gastroenterology. 1982;82:97–105.
25. Arroyo V, Ginès P, Rimola A, Gaya J. Renal-function abnormalities, prostaglandins, and effects of nonsteroidal anti-inflammatory drugs in cirrhosis with ascites – an overview with emphasis on pathogenesis. Am Med. 1986;81:104–22.
26. Salerno F, Badalamenti S. Drug induced renal failure in cirrhosis. In: Ginès P, Arroyo V, Rodés J, Schrier RW, editors. Ascites and Renal Dysfunction in Liver Diseases. Oxford: Blackwell, 2005:372–81.
27. Boyer TD, Zia P, Reynolds TB. Effect of indomethacin and prostaglandin A1 on renal function and plasma renin activity in alcoholic liver disease. Gastroenterology. 1979;77: 215–22.
28. Quintero E, Ginès P, Arroyo V et al. Sulindac reduces the urinary excretion of prostaglandins and impairs renal function in cirrhosis with ascites. Nephron. 1986;42: 298–303.
29. Claria J, Kent JD, Lopez-Parra M et al. Effects of celecoxib and naproxen on renal function in nonazotemic patients with cirrhosis and ascites. Hepatology. 2005;41:579–87.
30. Guevara M, Abecasis R, Terg R. Effect of celecoxib on renal function in cirrhotic patients with ascites. A pilot study. Scand J Gastroenterol. 2004;39:385–6.
31. Ginès P, Arroyo V, Quintero E et al. Comparison between paracentesis and diuretics in the treatment of cirrhotics with tense ascites. Results of a randomized study. Gastroenterology. 1987;93:234–41.
32. Hampel H, Bynum GD, Zamora E, El-Serag HB. Risk factors for the development of renal dysfunction in hospitalized patients with cirrhosis. Am J Gastroenterol. 2001;96:2206–10.
33. Cardenas A, Ginès P, Uriz J et al. Renal failure after upper gastrointestinal bleeding in cirrhosis: incidence, clinical course, predictive factors, and short-term prognosis. Hepatology. 2001;34:671–6.
34. Uriz J, Ginès P, Cárdenas A et al. Terlipressin plus albumin infusion: an effective and safe therapy of hepatorenal syndrome. J Hepatol. 2000;33:43–8.
35. Mulkay JP, Louis H, Donckier V et al. Long-term terlipressin administration improves renal function in cirrhotic patients with type 1 hepatorenal syndrome: a pilot study. Acta Gastroenterol Belg. 2001;64:15–19.
36. Moreau R, Durand F, Poynard T et al. Terlipressin in patients with cirrhosis and type 1 hepatorenal syndrome: a retrospective multicenter study. Gastroenterology. 2002;122:923–30.
37. Ortega R, Ginès P, Uriz J et al. Terlipressin therapy with and without albumin for patients with hepatorenal syndrome. Results of a prospective, non-randomized study. Hepatology. 2002;36:941–8.
38. Halimi C, Bonnard P, Bernard B et al. Effect of terlipressin (Glypressin) on hepatorenal syndrome in cirrhotic patients: results of a multicentre pilot study. Eur J Gastroenterol Hepatol. 2002;14:153–8.
39. Colle I, Durand F, Pessione F et al. Clinical course, predictive factors and prognosis in patients with cirrhosis and type 1 hepatorenal syndrome treated with terlipressin: a retrospective analysis. J Gastroenterol Hepatol. 2002;17:882–8.

40. Solanki P, Chawla A, Garg R et al. Beneficial effects of terlipressin in hepatorenal syndrome: a prospective, randomized placebo-controlled clinical trial. J Gastroenterol Hepatol. 2003;18:152–6.
41. Saner F, Kavuk I, Lang H et al. Terlipressin and gelafundin: safe therapy of hepatorenal syndrome. Eur J Med Res. 2004;9:78–82.
42. Saner FH, Fruhauf NR, Schafers RF et al. Terlipressin plus hydroxyethyl starch infusion: an effective treatment for hepatorenal syndrome. Eur J Gastroenterol Hepatol. 2003;15: 925–7.
43. Duvoux C, Zanditenas D, Hezode C et al. Effects of noradrenaline and albumin in patients with type 1 hepatorenal syndrome: a pilot study. Hepatology. 2002;36:374–80.
44. Angeli P, Guarda S, Fasolato S et al. Switch therapy with ciprofloxacin vs. intravenous ceftazidime in the treatment of spontaneous bacterial peritonitis in patients with cirrhosis: similar efficacy at lower cost. Aliment Pharmacol Ther. 2006;23:75–84.
45. Pomier-Layrargues G, Paquin SC, Hassoun Z, Lafortune M, Tran A. Octreotide in hepatorenal syndrome: a randomized, double-blind, placebo-controlled, crossover study. Hepatology. 2003;38:238–43
46. Angeli P, Volpin R, Gerunda G et al. Reversal of type 1 hepatorenal syndrome with the administration of midodrine and octreotide. Hepatology. 1999;29:1690–97.
47. Wong F, Pantea L, Sniderman K. Midodrine, octreotide, albumin, and TIPS in selected patients with cirrhosis and type 1 hepatorenal syndrome. Hepatology. 2004;40:55–64.
48. Esrailian E, Pantangco ER, Kyulo NL, Hu KQ, Runyon BA. Octreotide/midodrine therapy significantly improves renal function and 30-day survival in patients with type 1 hepatorenal syndrome. Dig Dis Sci. 2007 (Epub ahead of print).
49. Restuccia T, Ortega R, Guevara M et al. Effects of treatment of hepatorenal syndrome before transplantation on posttransplantation outcome. A case–control study. J Hepatol. 2004;40:140–6.
50. Testino G, Ferro C, Sumberaz A et al. Type-2 hepatorenal syndrome and refractory ascites: role of transjugular intrahepatic portosystemic stent–shunt in eighteen patients with advanced cirrhosis awaiting orthotopic liver transplantation. Hepatogastroenterology. 2003;50:1753–5.
51. Brensing KA, Textor J, Perz J et al. Long-term outcome after transjugular intrahepatic portosystemic stent–shunt in non-transplant cirrhotics with hepatorenal syndrome: a phase II study. Gut. 2000;47:288–95
52. Guevara M, Ginès P, Bandi JC et al. Transjugular intrahepatic portosystemic shunt in hepatorenal syndrome: effects on renal function and vasoactive systems. Hepatology. 1998; 28:416–22.

Section II
Liver diseases (Part 1)

Chair: PLM JANSEN and W STREMMEL

7
State-of-the-Art Lecture II: Hereditary haemochromatosis: the genes and the disease

A. PIETRANGELO

INTRODUCTION

The term haemochromatosis, or hereditary haemochromatosis, refers to a unique clinicopathological subset of iron-overload syndromes that are the result of a genetically determined failure to stop iron from entering the circulatory pool. The basis for this hereditary metabolic abnormality is inadequate hepcidin synthesis or activity. The subset currently includes the classic disorder related to HFE C282Y homozygosity and the rare disorders more recently attributed to loss of transferrin receptor 2 (TfR2), hepcidin (HAMP), or haemojuvelin (HJV) or to a specific subset of ferroportin FPN mutations[1] (Table 1).

MOLECULAR PATHOGENESIS

The first biochemical manifestation of haemochromatosis is an increase in transferrin saturation which reflects the uncontrolled influx of iron into the blood stream from enterocytes and macrophages that is normally down-regulated by hepcidin[2]. As discussed above, the iron-overload syndromes associated with *HFE*, *TfR2*, *HAMP*, and *HJV* mutations are all characterized by inadequate hepcidin synthesis[3–6]. Hepcidin expression in the liver is also significantly impaired in HFE-, TfR2- and HJV-knockout mice[7–9]. Therefore, hepcidin deficiency is the central pathogenic factor for all forms of haemochromatosis[10]. HJV, a bone-morphogenic protein co-receptor in the hepatocyte, is the key protein required for hepcidin transcription[11], whereas HFE and TfR2 may be involved in hepatocyte iron-sensing. The details of hepcidin regulation by HFE and TfR2 are still not defined. In hepatocytes from HFE- and TfR2-knockout mice, no role for bone morphogenic proteins (BMP) has been found[12], while an iron-modulated interaction between TfR2, HFE, and TfR1 has been proposed to convey the iron signal to hepcidin[13]. Discrete

Table 1 Main features of haemochromatosis and other hereditary iron overload disorders in humans

Disorder	*Affected gene (symbol/location)*	*Known or postulated gene product function*[a]	*Genetics*	*Mechanism for cellular iron accumulation*	*Clinical onset (decade)*	*Main clinical manifestation*	*Clinical course*
Haemochromatosis	Haemochromatosis gene (HFE/6p21.3)	Interaction with transferrin receptor 1 Uptake of transferrin-iron Hepcidin regulator	Autosomal recessive	Increased iron influx	3rd–5th	Liver disease	Mild–severe
	Transferrin-receptor 2 ((TfR2/7q22)	Uptake of transferrin and non-transferrin-bound iron Hepcidin regulator					
	Hepcidin antimicrobial peptide (HAMP/19q13.1)	Down-regulation of iron efflux from macrophages, enterocytes, placenta, throught degradation of ferroportin			2nd–3rd	Hypogadism and cardiac disease	Severe
	Haemojuvelin (HJV/1p21)	Co-receptor for bone morphogenic proteins Hepcidin transcriptional regulator					
	Solute carrier family 40 (iron-regulated transporter), membrane 1 (SLC40A1/2q32)	Iron export from cells including macrophages, enterocytes, placental cells	Autosomal dominant		3rd–5th	Liver disease	Unclear

Table 1 continued

Disorder	*Affected gene (symbol/location)*	*Known or postulated gene product function*[a]	*Genetics*	*Mechanism for cellular iron accumulation*	*Clinical onset (decade)*	*Main clinical manifestation*	*Clinical course*
Ferroportin disease	Solute carrier family 40 (iron-regulated transporter), member 1 (SLC40A1/2q32)	Iron export from cells including macrophages, enterocytes, placental cells	Autosomal dominant	Decreased iron efflux	4th–5th	Liver abnormalities Marginal anaemia	Mild
Acaeruloplasminaemia	Caeruloplasmin (CP/3q23-q25)	Iron efflux from cells	Autosomal recessive	Decreased iron efflux	2nd–3rd	Neurological manifestations Anaemia	Severe
A(hypo)transferrinaemia	Transferrin (Tf/3q21)	Iron transport in the blood stream	Autosomal recessive	Increased iron influx	1st–2nd	Anaemia	Severe

mutations in ferroportin have been reported to cause 'hepcidin resistance' *in vitro*[14–16], and they might cause an adult-onset form of iron loading in humans that is similar to *HFE* haemochromatosis[17,18].

Once iron saturates circulating transferrin, progressive accumulation in the parenchymal cells of key organs follows, creating a distinct risk for tissue iron overload and damage. The deposits are particularly evident in hepatocytes (Figure 1A) but may also be found in parenchymal cells of the endocrine glands and heart. This stage is reflected in increasing serum ferritin levels. The time of onset and pattern of organ involvement vary depending on the rate and

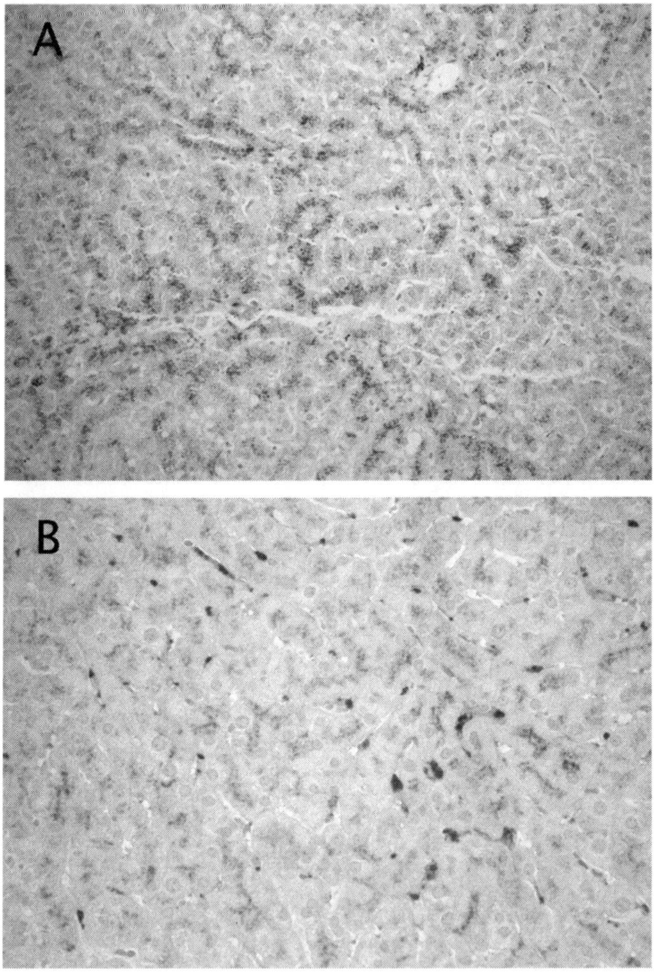

Figure 1 Iron accumulation patterns in haemochromatosis and ferroportin disease (Perls' Prussian blue stain). **A**: HFE haemochromatosis: iron deposits are evident in parenchymal cells, mainly in periportal areas; **B**: Ferroportin disease: iron is typically present in Kupffer cells and, to a lesser extent, hepatocytes

magnitude of plasma iron overloading, which are, in turn, related to the underlying mutation. This is the basis for the distinction between the milder adult-onset syndromes (*HFE*- and *TfR2*-related) and the more severe juvenile-onset forms (*HJV*- and *HAMP*-related)[10]. Iron is abundant (even excessive) in the plasma, and its transport to the bone marrow by transferrin and incorporation into haemoglobin are unaffected by the genetic defects that cause haemochromatosis. Consequently, anaemia is not a feature of this iron-overload syndrome, and therapeutic phlebotomy is tolerated quite well.

Classic ferroportin disease (i.e. that caused by mutations that reduce macrophage iron export) differs from haemochromatosis in fundamental respects[19]: the absence of the early expansion of the circulatory iron pool (transferrin saturation rates rise only in the later stages of this disease) and the presence of massive iron overloading of the reticuloendothelial macrophages, which typically present an iron-deprivation phenotype in haemochromatosis (Figure 1B). Abdominal magnetic resonance imaging (MRI) is thus a very useful, non-invasive tool for differentiating classic ferroportin disease, in which the ratio of splenic to bone iron signals is high (the 'macrophage-iron signature'), from haemochromatosis and from non-classic ferroportin disorders (i.e. those due to mutations that render ferroportin resistant to hepcidin degradation and result in unchecked iron export). Like haemochromatosis, these disorders are characterized on MRI by low spleen/bone iron signals[20]. The same type of 'compartmental analysis' reveals other possible pathophysiological subsets within the group of hereditary iron-overload syndromes.

CLINICAL ASPECTS

In adults, haemochromatosis can present with non-specific symptoms, simple biochemical iron abnormalities, or signs of severe organ disease. The most common presenting symptoms are fatigue, malaise, and arthralgia, and hepatomegaly is one of the earliest physical signs. Elevated serum transferrin saturation rates precede increases in serum ferritin. Transaminases may be normal or moderately increased. Increasing serum ferritin levels, rarely seen before the second–third decade, herald iron accumulation in tissues, and values above 1000 ng/ml may indicate underlying liver fibrosis in HFE haemochromatosis, even when transaminase levels are normal. The classical syndromic presentation (liver cirrhosis, bronze skin, diabetes and failure of other endocrine organs, or joint inflammation and heart disease) is less common today, thanks to screening and improved case detection due to greater clinician awareness of the disease. As discussed, most cases are the result of C282Y-HFE homozygosity, while they are much more rarely due to TfR2 mutations.

Individuals suffering from juvenile forms of haemochromatosis usually present with hypogonadism, which is inevitably found in those aged 20 or over. Cardiomyopathy and endocrine disorders, including diabetes, appear earlier than they do in adult-onset forms. This is probably due to the fact that the circulatory iron overload develops much more rapidly and reaches much higher levels (due to a more severe hepcidin deficiency associated with mutation of the 'juvenile-onset' genes). As a result, organs such as the heart or endocrine glands, which are more susceptible to iron toxicity, succumb to its effects earlier, and their failure will dominate the clinical picture. Usually the underlying pathogenic mutations involve *HJV* and only rarely *HAMP*.

It is important to recall that, variations notwithstanding, all of these mutations cause the same syndrome, the organs that can be damaged by iron toxicity are identical (i.e. liver, heart, endocrine glands, joints), and the pathogenic basis of all forms is hepcidin deficiency. Uncharacteristically severe or anticipated-onset disease in patients with mutations of an 'adult-onset gene' (e.g. HFE), might be the result of undetected mutations in other haemochromatosis juvenile genes such as *HAMP* or *HJV*[21,22] or even *TfR2*[6]. The variety of genotypes that can produce a haemochromatosis phenotype highlights the importance of defining and classifying this disease as a unique clinicopathological entity.

HFE-related haemochromatosis is the prototype form of haemochromatosis and by far the most common[10]. Two main HFE mutations exist, C282Y and H63D, but only C282Y homozygosity is capable of producing clinical manifestations. In the absence of comorbidity, compound C282Y/H63D heterozygosity has minimal clinical penetrance if any. Neither the H63D homozygosity nor compound heterozygosity for rarer HFE mutations, such as H65C, seems to confer a risk for organ disease[23–25]. The estimated prevalence of the C282Y mutation is high in Caucasians (1:200–300) and much lower in Hispanics, Asian Americans, Pacific Islanders, and black persons[26]. Studies of C282Y homozygotes have revealed elevated transferrin saturation rates in 71–86% of men and 40–100% of women homozygotes; serum ferritin elevations are found in 34–100% of men and 9–63% of women[27–32]. However, studies of unselected populations (i.e. enrolled regardless of symptoms or serum iron levels) indicate that the clinical penetrance of this genotype is much lower than previously thought[24]. As to the natural history, three uncontrolled longitudinal studies have not consistently identified increasing serum ferritin levels over time, and have not demonstrated an overt progression to clinical haemochromatosis[30,33,34]. Only 5% of patients identified through primary-care screening have cirrhosis[35]. Nearly 38% of C282Y homozygotes identified as a result of nonfamily-based genetic screening had iron overload, 25% liver fibrosis, and 6% cirrhosis[36]. Higher rates of iron overload (49–86%) and cirrhosis (8%) are found among the family members of probands[37,38]. In the Hemochromatosis and Iron Overload Screening (HEIRS) study, self-reported diabetes was not significantly associated with HFE mutations[39].

It is clear that the development of organ disease in haemochromatosis requires the concurrence of genetic and non-genetic factors[10]. Coinheritance of mutations in other haemochromatosis genes, such as *HAMP* and *HJV*, is also important[21,22]. The modifier effects of other iron-related genes have been documented in mice[40–42], but these findings have not been fully confirmed in humans[43–47]. There is evidence for a strong association between alcohol and development of haemochromatosis-related cirrhosis[48], while debate continues over the co-factorial roles of steatosis-fat and high BMI[49,50] and the effect of polymorphic changes in oxidative-stress-related genes[51].

The diagnosis is based on the presence of circulatory and tissue iron overload and C282Y homozygosity. The most widely used biochemical marker of iron overload is serum ferritin. In validation studies in which body iron stores were assessed by phlebotomy, serum ferritin proved to be a highly sensitive test for iron overload in haemochromatosis[52]. Although overload can be reliably excluded if ferritin levels are normal, the specificity of this marker is low: elevated values can be the result of various inflammatory or neoplastic states, diabetes and the metabolic syndrome, alcohol abuse, or viral hepatitis. Evaluation of the hepatic iron content and distribution by liver biopsy was the gold standard for diagnosis of haemochromatosis prior to the identification of HFE. Today the demonstration of HFE homozygosity in a subject with a high transferrin saturation rate and elevated serum ferritin level is sufficient for diagnosis of haemochromatosis[10], although liver biopsy can still play an important role in establishing the prognosis. In C282Y homozygotes (particularly those over 40 years of age), a serum ferritin level above 1000 ng/L, together with increased transaminases and hepatomegaly, may be an indication for liver biopsy to rule out hepatic fibrosis or cirrhosis[53–55].

Symptomatic subjects with clear signs of circulatory and tissue iron overload, but negative results in tests for HFE mutation, may have pathogenic mutations in other haemochromatosis genes. Genetic testing for non-HFE haemochromatosis is complex, however, and it is not widely available. An alternative approach in these cases is based on biopsy demonstration of the haemochromatotic pattern of hepatic iron distribution, together with a hepatic iron index[10] (the amount of iron in the liver – in micromoles – divided by age) of >1.9. The latter finding was considered diagnostic for HLA-linked haemochromatosis in the pre-HFE era.

THERAPY

Phlebotomy is the current standard of care in haemochromatosis. One unit (400–500 ml) of blood contains approximately 200–250 mg of iron. A weekly phlebotomy regimen can restore safe blood levels of iron (reflected by serum ferritin levels of less than 20–50 μg/L and transferrin saturation below 30%) in 1–2 years' time. Maintenance therapy, which typically involves removal of 2–4 units a year, can then be initiated. This must be continued for the duration of the patient's life to keep transferrin saturation and ferritin normal. Phlebotomy is generally believed to have few side-effects and to be effective in removing iron from tissues and preventing complications. However, these views have never been confirmed in a randomized, controlled trial, and no studies have compared the effectiveness of early and delayed onset of treatment. In published case series[56–58], the 10-year survival of individuals recently diagnosed with haemochromatosis or treated prior to the development of cirrhosis did not differ from that in age- and sex-matched population controls. In another study 158 Danish patients were followed for a median period of 8.5 years, and survival of those who were adequately phlebotomized was significantly greater (estimated Kaplan–Meier survival rates at 5 years: 93% versus 48% among those who were not adequately phlebotomized;

corresponding rates at 10 years were 78% versus 32%)[59]. Furthermore, even among patients with cirrhosis or diabetes, those who were adequately treated had better survival than those who were not. When the effect of treatment was evaluated on disease progression in individuals with biopsy-proven liver fibrosis, phlebotomy was associated with improvement of 13–50%; the highest rates were observed among individuals with the mildest degrees of liver fibrosis at baseline[56,60,61]. The hypogonadism, cirrhosis, destructive arthritis, and insulin-dependent diabetes associated with haemochromatosis are usually irreversible. Several studies, however, suggest that phlebotomy can improve certain aspects of these diseases. Reductions have been reported in daily insulin doses in type 1 diabetes mellitus, aminotransferase levels, weakness, lethargy, and abdominal pain[36]. If phlebotomy is contraindicated, or not tolerated, other iron removal strategies such as the use of iron-chelators can be considered.

CONCLUSIONS AND PERSPECTIVES

Haemochromatosis was originally considered an unusual autopsy finding that was probably alcohol-related. Over a century later it was finally recognized as a hereditary disorder caused by mutation of an HLA-linked protein thought to be involved in intestinal iron transport. Soon after the identification of HFE as 'the cause' of the disease, other apparently unrelated haemochromatosis proteins emerged. In addition, it gradually became clear that HFE played no direct role in intestinal iron transport. The definition of haemochromatosis and its pathogenesis has become increasingly confusing, particularly for clinicians. As we recognize it today, haemochromatosis results from the non-linear interaction of pathogenic mutations in a variety of genes involved in hepcidin synthesis/activity and environmental factors, such as alcohol. The dissection of the pathogenic pathways involved in the disrupted hepcidin production and iron homeostasis will allow us to develop new and more effective tools for diagnosis and management of this disorder.

References

1. Pietrangelo A. Hereditary hemochromatosis. Annu Rev Nutr. 2006;26:251–70.
2. Nemeth E, Tuttle MS, Powelson J et al. Hepcidin regulates cellular iron efflux by binding to ferroportin and inducing its internalization. Science. 2004;306:2090–3.
3. Gehrke SG, Kulaksiz H, Herrmann T et al. Expression of hepcidin in hereditary hemochromatosis: evidence for a regulation in response to serum transferrin saturation and non-transferrin-bound iron. Blood. 2003;102:371–6.
4. Nemeth E, Roetto A, Garozzo G, Ganz T, Camaschella C. Hepcidin is decreased in TFR2 hemochromatosis. Blood. 2005;105:1803–6.
5. Papanikolaou G, Samuels ME, Ludwig EH et al. Mutations in HFE2 cause iron overload in chromosome 1q-linked juvenile hemochromatosis. Nat Genet. 2004;36:77–82.
6. Pietrangelo A, Caleffi A, Henrion J et al. Juvenile hemochromatosis associated with pathogenic mutations of adult hemochromatosis genes. Gastroenterology. 2005;128:470–9.
7. Muckenthaler M, Roy CN, Custodio AO et al. Regulatory defects in liver and intestine implicate abnormal hepcidin and Cybrd1 expression in mouse hemochromatosis. Nat Genet. 2003;34:102–7.

8. Kawabata H, Fleming RE, Gui D et al. Expression of hepcidin is down-regulated in TfR2 mutant mice manifesting a phenotype of hereditary hemochromatosis. Blood. 2005;105: 376–81.
9. Huang FW, Pinkus JL, Pinkus GS, Fleming MD, Andrews NC. A mouse model of juvenile hemochromatosis. J Clin Invest. 2005;115:2187–91.
10. Pietrangelo A. Hereditary hemochromatosis – a new look at an old disease. N Engl J Med. 2004;350:2383–97.
11. Babitt JL, Huang FW, Wrighting DM et al. Bone morphogenetic protein signaling by hemojuvelin regulates hepcidin expression. Nat Genet. 2006;38:531–9.
12. Truksa J, Peng H, Lee P, Beutler E. Bone morphogenetic proteins 2, 4, and 9 stimulate murine hepcidin 1 expression independently of Hfe, transferrin receptor 2 (Tfr2), and IL-6. Proc Natl Acad Sci USA. 2006;103:10289–93.
13. Goswami T, Andrews NC. Hereditary hemochromatosis protein, HFE, interaction with transferrin receptor 2 suggests a molecular mechanism for mammalian iron sensing. J Biol Chem. 2006;281:28494–8.
14. Drakesmith H, Schimanski LM, Ormerod E et al. Resistance to hepcidin is conferred by hemochromatosis-associated mutations of ferroportin. Blood. 2005;106:1092–7.
15. Schimanski LM, Drakesmith H, Merryweather-Clarke AT et al. *In vitro* functional analysis of human ferroportin (FPN) and hemochromatosis-associated FPN mutations. Blood. 2005;105:4096–102.
16. De Domenico I, Ward DM, Nemeth E et al. The molecular basis of ferroportin-linked hemochromatosis. Proc Natl Acad Sci USA. 2005;102:8955–60.
17. Wallace DF, Clark RM, Harley HA, Subramaniam VN. Autosomal dominant iron overload due to a novel mutation of ferroportin1 associated with parenchymal iron loading and cirrhosis. J Hepatol. 2004;40:710–13.
18. Sham RL, Phatak PD, West C, Lee P, Andrews C, Beutler E. Autosomal dominant hereditary hemochromatosis associated with a novel ferroportin mutation and unique clinical features. Blood Cells Mol Dis. 2005;34:157–61.
19. Pietrangelo A. The ferroportin disease. Blood Cells Mol Dis. 2004;32:131–8.
20. Pietrangelo A, Corradini E, Ferrara F et al. Magnetic resonance imaging to identify classic and nonclassic forms of ferroportin disease. Blood Cells Mol Dis. 2006;37:192–6.
21. Jacolot S, Le Gac G, Scotet V, Quere I, Mura C, Ferec C. HAMP as a modifier gene that increases the phenotypic expression of the HFE pC282Y homozygous genotype. Blood. 2004;103:2835–40.
22. Le Gac G, Scotet V, Ka C et al. The recently identified type 2A juvenile haemochromatosis gene (HJV), a second candidate modifier of the C282Y homozygous phenotype. Hum Mol Genet. 2004;13:1913–18.
23. Gochee PA, Powell LW, Cullen DJ, Du Sart D, Rossi E, Olynyk JK. A population-based study of the biochemical and clinical expression of the H63D hemochromatosis mutation. Gastroenterology. 2002;122:646–51.
24. Waalen J, Nordestgaard BG, Beutler E. The penetrance of hereditary hemochromatosis. Best Pract Res Clin Haematol. 2005;18:203–20.
25. Samarasena J, Winsor W, Lush R, Duggan P, Xie Y, Borgaonkar M. Individuals homozygous for the H63D mutation have significantly elevated iron indexes. Dig Dis Sci. 2006;51:803–7.
26. Adams PC, Reboussin DM, Barton JC et al. Hemochromatosis and iron-overload screening in a racially diverse population. N Engl J Med. 2005;352:1769–78.
27. Beutler E, Felitti VJ, Koziol JA, Ho NJ, Gelbart T. Penetrance of 845G→A (C282Y) HFE hereditary haemochromatosis mutation in the USA. Lancet. 2002;359:211–18.
28. Jackson HA, Carter K, Darke C et al. HFE mutations, iron deficiency and overload in 10,500 blood donors. Br J Haematol. 2001;114:474–84.
29. Deugnier Y, Jouanolle AM, Chaperon J et al. Gender-specific phenotypic expression and screening strategies in C282Y-linked haemochromatosis: a study of 9396 French people. Br J Haematol. 2002;118:1170–8.
30. Andersen RV, Tybjaerg-Hansen A, Appleyard M, Birgens H, Nordestgaard BG. Hemochromatosis mutations in the general population: iron overload progression rate. Blood. 2004;103:2914–19.

31. Chambers V, Sutherland L, Palmer K et al. Haemochromatosis-associated HFE genotypes in English blood donors: age-related frequency and biochemical expression. J Hepatol. 2003;39:925–31.
32. Olynyk JK, Cullen DJ, Aquilia S, Rossi E, Summerville L, Powell LW. A population-based study of the clinical expression of the hemochromatosis gene [see comments]. N Engl J Med. 1999;341:718-24.
33. McLaren CE, Barton JC, Adams PC et al. Hemochromatosis and Iron Overload Screening (HEIRS) study design for an evaluation of 100,000 primary care-based adults. Am J Med Sci. 2003;325:53–62.
34. Olynyk JK, Hagan SE, Cullen DJ, Beilby J, Whittall DE. Evolution of untreated hereditary hemochromatosis in the Busselton population: a 17-year study. Mayo Clin Proc. 2004;79: 309–13.
35. Schmitt B, Golub RM, Green R. Screening primary care patients for hereditary hemochromatosis with transferrin saturation and serum ferritin level: systematic review for the American College of Physicians. Ann Intern Med. 2005;143:522–36.
36. Whitlock EP, Garlitz BA, Harris EL, Beil TL, Smith PR. Screening for hereditary hemochromatosis: a systematic review for the U.S. Preventive Services Task Force. Ann Intern Med. 2006;145:209–23.
37. Gleeson F, Ryan E, Barrett S, Crowe J. Clinical expression of haemochromatosis in Irish C282Y homozygotes identified through family screening. Eur J Gastroenterol Hepatol. 2004;16:859–63.
38. McCune CA, Ravine D, Carter K et al. Iron loading and morbidity among relatives of HFE C282Y homozygotes identified either by population genetic testing or presenting as patients. Gut. 2006;55:554–62.
39. Acton RT, Barton JC, Passmore LV et al. Relationships of serum ferritin, transferrin saturation, and HFE mutations and self-reported diabetes in the Hemochromatosis and Iron Overload Screening (HEIRS) study. Diabetes Care. 2006;29:2084–9.
40. Levy JE, Montross LK, Andrews NC. Genes that modify the hemochromatosis phenotype in mice. J Clin Invest. 2000;105:1209–16.
41. Bensaid M, Fruchon S, Mazeres C, Bahram S, Roth MP, Coppin H. Multigenic control of hepatic iron loading in a murine model of hemochromatosis. Gastroenterology. 2004;126: 1400–8.
42. Tolosano E, Fagoonee S, Garuti C et al. Haptoglobin modifies the hemochromatosis phenotype in mice. Blood. 2005;105:3353–5.
43. Lee PL, Gelbart T, West C, Halloran C, Felitti V, Beutler E. A study of genes that may modulate the expression of hereditary hemochromatosis: transferrin receptor-1, ferroportin, ceruloplasmin, ferritin light and heavy chains, iron regulatory proteins (IRP)-1 and -2, and hepcidin. Blood Cells Mol Dis. 2001;27:783–802.
44. Lee P, Gelbart T, West C, Halloran C, Beutler E. Seeking candidate mutations that affect iron homeostasis. Blood Cells Mol Dis. 2002;29:471–87.
45. Van Vlierberghe H, Langlois M, Delanghe J et al. Haptoglobin phenotype 2-2 overrepresentation in Cys282Tyr hemochromatotic patients. J Hepatol. 2001;35:707–11.
46. Carter K, Bowen DJ, McCune CA, Worwood M. Haptoglobin type neither influences iron accumulation in normal subjects nor predicts clinical presentation in HFE C282Y haemochromatosis: phenotype and genotype analysis. Br J Haematol. 2003;122:326–32.
47. Kelleher T, Ryan E, Barrett S, O'Keane C, Crowe J. DMT1 genetic variability is not responsible for phenotype variability in hereditary hemochromatosis. Blood Cells Mol Dis. 2004;33:35–9.
48. Fletcher LM, Dixon JL, Purdie DM, Powell LW, Crawford DH. Excess alcohol greatly increases the prevalence of cirrhosis in hereditary hemochromatosis. Gastroenterology. 2002;122:281–9.
49. Powell EE, Ali A, Clouston AD et al. Steatosis is a cofactor in liver injury in hemochromatosis. Gastroenterology. 2005;129:1937–43.
50. Adams LA, Angulo P, Abraham SC, Torgerson H, Brandhagen D. The effect of the metabolic syndrome, hepatic steatosis and steatohepatitis on liver fibrosis in hereditary hemochromatosis. Liver Int. 2006;26:298.-304.
51. Osterreicher CH, Datz C, Stickel F et al. TGF-beta1 codon 25 gene polymorphism is associated with cirrhosis in patients with hereditary hemochromatosis. Cytokine. 2005;31: 142–8.

52. Beutler E, Felitti V, Ho NJ, Gelbart T. Relationship of body iron stores to levels of serum ferritin, serum iron, unsaturated iron binding capacity and transferrin saturation in patients with iron storage disease. Acta Haematol. 2002;107:145–9.
53. Guyader D, Jacquelinet C, Moirand R et al. Noninvasive prediction of fibrosis in C282Y homozygous hemochromatosis. Gastroenterology. 1998;115:929–36.
54. Beaton M, Guyader D, Deugnier Y, Moirand R, Chakrabarti S, Adams P. Noninvasive prediction of cirrhosis in C282Y-linked hemochromatosis. Hepatology. 2002;36:673–8.
55. Morrison ED, Brandhagen DJ, Phatak PD et al. Serum ferritin level predicts advanced hepatic fibrosis among US patients with phenotypic hemochromatosis. Ann Intern Med. 2003;138:627–33.
56. Niederau C, Fischer R, Purschel A, Stremmel W, Haussinger D, Strohmeyer G. Long-term survival in patients with hereditary hemochromatosis [see comments]. Gastroenterology. 1996;110:1107–19.
57. Adams PC, Speechley M, Kertesz AE. Long-term survival analysis in hereditary hemochromatosis. Gastroenterology. 1991;101:368–72.
58. Bomford A, Williams R. Long term results of venesection therapy in idiopathic haemochromatosis. Q J Med. 1976;45:611–23.
59. Milman N, Pedersen P, Steig T, Byg KE, Graudal N, Fenger K. Clinically overt hereditary hemochromatosis in Denmark 1948–1985: epidemiology, factors of significance for long-term survival, and causes of death in 179 patients. Ann Hematol. 2001;80:737–44.
60. Powell LW, Dixon JL, Ramm GA et al. Screening for hemochromatosis in asymptomatic subjects with or without a family history. Arch Intern Med. 2006;166:294–301.
61. Falize L, Guillygomarc'h A, Perrin M et al. Reversibility of hepatic fibrosis in treated genetic hemochromatosis: a study of 36 cases. Hepatology. 2006;44:472–7.

8
Wilson disease: practical applications of molecular approaches

D. W. COX

INTRODUCTION

Wilson disease (WND) has been recognized as a disease entity for more than 100 years. However, the past decade has brought major advances both in our understanding and in improved diagnosis of this condition. Earliest descriptions of 'pseudosclerosis' by C. Westphal in 1883 and A. Strumpell in 1898, in Germany, were recognized later as a neurological form of WND. S.A. Kinnear Wilson, in 1912, described a progressive lenticular degeneration, affecting both brain and liver[1]. In 1948, B.M. Mandelbrote and J.N. Cumings recognized WND as a disorder of copper accumulation.

Caeruloplasmin is a copper-containing plasma protein that is often decreased in WND. At one time caeruloplasmin was considered as possibly having no important function, perhaps being only a 'vestigial' protein. Its role as an important ferroxidase required for the transport of iron out of cells was recognized after identification of the recessive disorder, acaeruloplasminaemia, in which iron accumulation causes neurological and retinal deterioration, and diabetes[2,3]. In addition to the important role of caeruloplasmin in iron transport, other proteins that require copper for their function include cytochrome oxidase (electron transfer), lysyl oxidase (connective tissue and elastin crosslinking), superoxide dismutase (free radical scavenging), dopamine β-mono-oxygenase (neurotransmission) and tyrosinase (pigment production). The average human intake of copper is approximately 1.5–3 mg copper per day. Copper is absorbed in the upper intestine and is transferred to plasma where it is bound to albumin. Albumin and copper histidine transport the copper to tissues, with the majority being transported to the liver. ATOX1, one of the class of low molecular weight metallochaperones within the hepatocyte, delivers copper to ATP7B[4].

While copper is essential for these enzymes most is eliminated from the liver, mediated by ATP7B. When this membrane transporter is defective, copper accumulates in tissues, particularly liver and brain, leading to production of free radicals and to tissue damage. This abundance of free radicals is reflected in the low serum levels of the antioxidant vitamin E in some WND patients[5,6].

FEATURES OF WND

The diagnosis of WND has typically been based on its clinical and biochemical features. However, gene discovery, and subsequent mutation analysis, is demonstrating that WND remains a diagnostic challenge, and can easily be missed, because of the great variability in a features of the disease. Prompt diagnosis is essential since treatment is reasonably simple and effective, using chelation or oral zinc, before tissue damage occurs. WND is an autosomal recessive disorder, occurring in many populations with an incidence of about 1 in 30 000. In China and Sardinia, among others, the frequency is as high as 1 in 10 000. Recent reviews provide details of the clinical and biochemical features and further references[7–10].

Clinical features

Clinical features are highly variable, contributing to the difficulty of diagnosis. Hepatic disease can be the main form of presentation, and can be acute or chronic. Neurological manifestations include tremor and/or rigidity[11]. Patients can have hepatic or neurological features only, or a combination of both. Psychiatric disorders, kidney disease and haemolysis can be manifestations of WND. Presentation as haemolytic anaemia is an infrequent mode of first presentation, but is also one that is easily missed. Haemolysis occurs as a result of oxidative stress associated with increased non-caeruloplasmin-bound copper in the blood[12]. The onset of disease symptoms can occur as early as 3 years of age[13], and has been reported into the seventh decade[14,15]. Most commonly, onset of symptoms is between 8 and 50 years of age. The wide range of clinical features and age of onset contribute to the diagnostic challenge of WND. Mutation analysis can be a valuable aid to ensure the correct diagnosis.

Kayser–Fleischer (KF) rings

B. Kayser and B. Fleischer, respectively in 1902 and 1903, noted brownish rings on the outer circumference of the cornea that we now recognize as copper deposition. KF rings, considered to be a characteristic feature of WND, are often absent.

Biochemical features

Biochemical assays reflecting copper status have traditionally been used for diagnosis; however, none is specific[16].

Serum caeruloplasmin

Serum caeruloplasmin, a copper-containing ferroxidase, is present at a very low concentration at birth, peaking at about 3 years of age, to 300–500 mg/L), gradually declining to adult normal levels of 200–400 mg/L[17]. Caeruloplasmin, as an acute-phase protein, is elevated by pregnancy, oestrogen administration, inflammation and liver disease. Caeruloplasmin is often used as the first biochemical assay for patients with clinical signs and symptoms suggestive of WND. In patients confirmed by mutation analysis to have WND, borderline or normal caeruloplasmin was identified in 20% of cases with a neurological onset, and 30% of cases with hepatic onset. Decreased serum caeruloplasmin is not a reliable diagnostic sign, particularly for those with a predominantly hepatic onset[18]. The non-caeruloplasmin component of serum increases as body copper stores increase, but may be scarcely elevated in early phases of the disease. Caeruloplasmin is measured immunologically in many laboratories, but the enzymatic method is preferable, measuring only that component that contains copper[19,20]. This is especially important in patients on treatment with chelation therapy of many years duration, as absence of the enzymatic component suggests copper depletion.

Urinary copper

In WND patients, 24-h urinary copper is usually increased well above the normal limit of 0.6 μmol/24 h, but may be borderline in patients in the early phase of the disease.

Hepatic copper

WND is usually associated with a hepatic copper concentration of >250 μg/g dry weight liver. However, the concentration may be only slightly elevated in patients with neurological disease. Furthermore, copper is not distributed evenly throughout the liver and a biopsy needle can, by chance, sample a region with little excess copper[21]. This will provide an unreliable estimate of total liver copper.

The occurrence of the above features in patients tested in our laboratory is shown in Table 1. Patients are divided according to the predominant type of onset: neurological or hepatic. Confirmed patients are those in whom biochemical and clinical features led to a clear diagnosis of WND, and were part of a study of the mutation spectrum in known patients. 'Query' patients are those in which the diagnosis was not clear based on clinical and biochemical features expected for WND.

Table 1 Traditional diagnostic features of Wilson disease

	Hepatic onset		*Neurological onset*	
	Confirmed (%)	*Query (%)*	*Confirmed (%)*	*Query (%)*
K-F rings	89	50	81	69
Low Cp	90	65	100	75
Urinary Cu >1.6 μM	100	90	100	90

Another diagnostic problem is that heterozygotes sometimes have features that overlap with those of patients[22], causing some heterozygotes to be treated as patients. They can have a markedly decreased serum caeruloplasmin, borderline or slightly increased urinary copper, and borderline or slightly increased hepatic copper. In spite of such copper abnormalities, heterozygotes are not known to develop clinical symptoms or to need treatment. Heterozygotes occur at a frequency of about 1% in most populations, as calculated from the disease frequency.

THE GENE DEFECT IN WILSON DISEASE

In 1993, discovery of the causative gene for WND, ATP7B, was an exciting event in our laboratory, and has provided a major breakthrough for the understanding of WND and copper transport mechanisms[23,24]. In addition, the opportunity arose for reliable molecular diagnosis by mutation analysis. The *ATP7B* gene encodes a copper-transporting ATPase, ATP7B, and is expressed in liver, kidney and brain. The ATP7B transporter excretes copper from cells of these organs. Expression is highest in the liver, and ATP7B is a key membrane transporter for excreting copper into the bile. Mammalian ATP7B has an N-terminal series of six copper-binding domains forming a secondary structure. Coppers bound to the CXXC (cysteine) motifs in the domain bind ATOX1, the low molecular weight copper chaperone that initiates the transfer of copper through the Golgi membrane. There are eight trans-membrane (Tm) domains. A CPC motif in Tm 6 provides specificity for copper transport. A TEG motif is required for the transduction of energy. A phosphorylated aspartate residue in the DKTG motif is typical of P-type ATPases[23]. These features are common to metal transporting ATPases, particularly the calcium transporter SERCA1, with a similar structural conformation.

Menkes disease is another disorder of copper transport, caused by a defect in the *ATP7A* gene, located on the X chromosome. The gene has a high degree of sequence similarity to that of *ATP7B*, but the expression of *ATP7A* is very different, occurring in many tissues, but particularly in the intestine. A defect in ATP7A results in lack of absorption of copper from the intestine, and subsequent widespread copper deficiency.

THE SPECTRUM OF MUTATIONS

Mutations occur throughout the length of the *ATP7B* gene, and number more than 290, identified since our first mutation enumerations in 1993 and 1995[23,25]. Mutations in the promoter region are relatively rare. There are many approaches to mutation detection. Single-strand conformation polymorphism (SSCP) analysis has been used in many populations. Other methods of mutation analysis are available. However, development of rapid sequencing methods makes mutation detection by sequencing feasible and cost-effective, especially considering the high cost of disability from delayed or missed diagnosis of this treatable condition. The distribution of various types

Table 2 Mutations in the Wilson and Menkes disease genes[a]

	Wilson *(ATP7B) (%)*	*Menkes*[b] *(ATP7A) (%)*
Deletion/insertion	27	39
Nonsense	7	21
Missense	58	17
Splice site	7	23

[a]Data from ref. 26.

[b]Large Menkes deletions (20%) are excluded.

of mutation: missense, nonsense, small insertions and deletions, and splice site mutations, are shown in Table 2. Mutations are listed, with references for each, in the database maintained at the University of Alberta, Department of Medical Genetics: http://www.medicalgenetics.med.ualberta.ca/wilson/index.php. The database also includes information on the likelihood that the mutation is disease-causing, based on type of mutation, absence in ethnically matched controls, and functional analysis. The database currently includes 292 mutations: 168 missense mutations, 60 small deletions and 21 insertions, 20 nonsense mutations and 23 splice site mutations. Insertions, deletions, and nonsense mutations occur throughout the gene. Missense mutations are more highly focused in the important functional regions.

In spite of the high degree of sequence similarity between *ATP7A* and *ATP7B*, the mutation spectrum is very different[26], as shown in Table 2. Large deletions, involving one or more exons, occur in 15–20% of chromosomes in Menkes disease patients, but are rare in WND. Only one large deletion, of exons 20 and 21, has been reported in WND[27]. The reason for this difference is not clear. In our laboratory we have tested for the occurrence of large deletions, using multiplex ligation-dependent probe amplification (MLPA) (MRC-Holland). This method can be used when sequencing identifies no mutations, but biochemical features are supportive of a diagnosis of WND. Of 32 potential WND chromosomes we tested, none was found to have large deletions of *ATP7B* (Cox et al., unpublished).

We are currently using VariantSEQ*r*™ (Applied Biosystems, Foster City, CA), with a sequential protocol for sequencing all exons if needed. Analysis is carried out with SeqScape® Software. We have been able to obtain results in 3 days if needed in an emergency. Technicalities of mutation detection are no longer a deterrent to mutation testing.

However, a current problem lies in deciding which mutations are disease-causing and which are rare normal variants. Most mutations identified in *ATP7B* are missense mutations. Small deletions and insertions usually result in a frameshift that is predicted to disrupt gene function. Nonsense mutations cause premature truncation of the protein. Splice site mutations occur throughout the gene and are predicted to usually produce a non-functional protein. In Menkes disease, splice site mutations frequently show a more mild disease, occipital horn syndrome[28,29].

Mutations in specific populations

Specific mutation types and frequencies are characteristic for specific ethnic groups. Very few mutations in ATP7B are common. A large number of mutations are found, each at a low frequency, in populations of European origin. The exception is the histidine1069glutamine (H1069Q) mutation[24] in about 35–45% in mixed European populations, and even higher in Eastern Europe[30]. The arginine778leucine (R778L) mutation[25] is a frequent mutation in Asian populations, accounting for over 50% of WND alleles in these populations. A 15 base pair deletion in the promoter region is common in the Sardinian population[31]. Interestingly, this is the only disease mutation identified to date in the promoter region[32]. Knowledge of which specific mutations occur in specific populations can be used diagnostically with low-cost methods in some populations, for example in Eastern Germany[33], Sardinia[34] and for the common leucine708proline (L708P) mutation almost unique to the Canary Islands. Even with complete sequencing of all exons, there are about 2% of patients in whom we cannot identify any mutation (Cox, unpublished). No doubt, in rare cases, a mutation lies within an intron, or even in flanking sequence at some currently unidentified site important for function. Examination of the RNA could be useful for detection of intronic splice defects. Another possibility is the presence of other defects in copper transport, not yet identified.

Importance of functional assessment of mutations

A large number of missense mutations, of which less than 20% have been tested for function, now pose the greatest problem for molecular diagnosis. For small deletions and insertions causing frame shifts, or nonsense mutations, function of the protein can be predicted to be disrupted function of ATP7B in most cases[25,35], except with the occurrence of exon skipping to produce a functional protein. Splice site mutations cannot be assumed to affect splicing in all cases, and need to be functionally tested[36]. While the technicalities of rapid mutation detecting are no longer an issue, discrimination of disease variants from rare normal variants is now the greatest problem for accurate diagnosis of WND.

ATP7B has two essential functions within the cell: (1) a transport function, in which copper is incorporated into caeruloplasmin and transported through the cell membrane; and (2) a trafficking function, by which ATP7B is cycled within the cell[37]. ATP7B traffics from the trans-Golgi network/late endosome compartment, to cytoplasmic vesicles, then to the apical membrane of the hepatocyte and other polarized cells[38,39]. An assay in yeast is useful for assessing the transport function[40–42]. The yeast *Saccharomyces cerivisiae* has a copper transport system similar to that of humans and other higher organisms. *CCC2*, the *ATP7B* counterpart in yeast, *can* be knocked out and a normal or variant human gene substituted. FET3 is the yeast counterpart of caeruloplasmin. Caeruloplasmin and FET3 are each required for cellular iron transport, in mammals and yeast, respectively. Wild-type yeast, yeast with a normal human *ATP7B*, or mutant *ATP7B* is then grown in appropriate media to test transport function. Yeasts with a functioning *CCC2* gene do not require the addition of iron and copper to their media, while a defective mutant has this requirement. The trafficking function can also be assessed experimentally. We, and others, have used various cells in culture, to examine the ability of mutant ATP7B to traffic within the cell, and to observe its resulting cellular distribution[41,43–45]. In our laboratory we have successfully used CHO cells for these assays. Mutants tested for functional activity are indicated in the Wilson Disease Database.

We have developed a toxicity assay, in CHO cells, that can be used in a microtitre plate to test for abnormal function (Cox, unpublished). Mutants shown to be defective can subsequently be tested for trafficking function using the same cells and constructs.

In addition to functional assessment, we compare the properties of the mutant ATP7B with those of normal ATP7B. These properties include hydrophobicity, size, and conservation of the amino acid residue in the ATP7B or Menkes protein (ATP7A) in other species. We have recently reported 24 new mutations with comparison of these properties in the mutant with the normal ATP7B[46].

Structural models for ATP7B have been proposed[47–49]. We use a three-dimensional model based on protein structure of the calcium transporter, SERCA1, which has been crystallized, to model the functional regions, to help assess the effects of mutations. Analysis software is Modbase[50], found at http://modbase.compbio.ucsf.edu/modbase-cgi/.

DIAGNOSTIC ISSUES AND PITFALLS

When characteristic clinical features and at least some abnormalities of copper biochemistry are present, the identification of one of the two gene mutations may be adequate. However, this one mutation must be well established as a mutation with no uncertainty regarding its status (disease-causing versus normal). For a patient with atypical clinical features, or asymptomatic, the possibility of heterozygote (carrier) status needs to be considered. The identification of both mutations is always preferred.

Discovery of *ATP7B* as the gene responsible for WND has made possible the accurate diagnosis of this highly variable disease. The identification of mutations has demonstrated that the typical clinical and biochemical features are not always reliable for the diagnosis. Accurate and timely diagnosis should help to avoid the long-term disability and death that can result from late or no treatment. The potential diagnosis of WND needs to be kept in mind for patients with any of the hepatic common neurological, psychiatric or haematological features suggestive of this disease. Technically, the mutations can be readily detected. However, functional studies are urgently needed to ensure the best use of the new mutation detection technologies.

DIAGNOSIS OF SIBS OF PATIENTS

Molecular approaches are invaluable, and essential, for the correct diagnosis of asymptomatic sibs of WND patients with the secure diagnosis of WND. In some families heterozygotes cannot be distinguished from affected homozygotes[22,51]. Heterozygotes can have low serum caeruloplasmin, borderline increased urinary copper, and increased hepatic copper. When both ATP7B mutations have been identified in the patient, sibs can be tested by direct mutation analysis for the known mutations. However, where the diagnosis is secure based on clinical and biochemical findings (an essential prerequisite), marker analysis can be used reliably.

The use of variable markers, such as dinucleotide repeat markers, or microsatellites, is important in all cases for correct diagnosis of sibs, where mutations in a known patient have not been analysed[52]. Accurate diagnosis by haplotype analysis requires confidence that the copper storage in the initial patient is in fact due to WND. A group of single-nucleotide markers within the gene can also be used. Both approaches can now be done easily and rapidly with current technologies. Marker studies are essential, particularly because of the overlap in biochemical features between presymptomatic patients and heterozygotes. In families in which molecular testing has not been used to confirm the genetic status, heterozygotes may be on lifelong treatment, with possible consequences of copper depletion. Appropriate copper analyses can be carried out to confirm the diagnosis of presymptomatic patients as made by marker studies. In populations in which there is a possibility for distant relationships, as in geographical isolates or in communities that have originated from a small number of ancestors, carrier testing in the family is recommended. We have observed a patient in whom her relatively rare presentation as haemolytic anaemia was recognized as WND only because testing had been carried out in the grandmother in a large previous family study (Cox, unpublished).

OTHER COPPER STORAGE DISORDERS

No other inherited primary disorders of copper storage have been identified to date. One such possibility arose through the study of a copper storage disease in dogs. Bedlington terriers develop a copper storage disorder, copper toxicosis, that can lead to liver disease. Through linkage analysis, copper toxicosis was identified as linked to a specific marker, C0147, on canine chromosome 10[53]. Further linkage studies led to the identification of a deletion of an exon in a gene MURR1 (now called COMMD1) as a potential causative gene for this disease[54,55]. However, while we have found that most affected dogs are homozygous for this deletion, at least some are heterozygous or do not carry the deletion at all[55], an observation also reported in Australian Bedlington terriers[56].

MURR1/COMMD1 interacts with the amino terminus of ATP7B, and could therefore be involved in the transport function[57]. MURR1 has been reported to have several functions, including: regulation of the sodium channel that regulates blood pressure and salt balance through interaction with human delta epithelial sodium channel[58], interaction with XIAP, an X-linked suppressor of apoptosis[59], and restriction of HIV-1 replication in lymphocytes through inhibition of NF-κB activity[60]. However, evidence is mounting that that this is unlikely to be the primary cause of Bedlington copper toxicosis, although it may modify the response to copper storage.

An important question is whether MURR1 could be responsible for copper storage disease in humans. In a study examining 24 children with copper storage associated with liver disease early in childhood[61], no mutations in the MURR1 gene were identified. We have sequenced MURR1 in patients from ages 6–66 years that show some evidence of copper storage, in some cases at least in part suggestive of WND[52]. Again, no disease-related mutations were found, but two common polymorphisms were identified. One of these polymorphisms has been suggested to modify the course of WND[62]. No effect of MURR1 as a modifier was observed in other studies[63–65].

CONCLUSION

Accurate and early diagnosis by DNA technologies is possible as the result of identification of ATP7B as the defective gene in WND. DNA diagnosis is essential for sibs of known patients and does not require the identification of both mutations in the patient. Our challenge now is to ensure the accuracy of molecular diagnostic approaches to identify all of the functionally important changes in ATP7B.

A challenge to the medical community is to identify all cases of Wilson disease so that appropriate treatment can be started in a timely manner to prevent early death or lifelong disability.

Acknowledgements

Mutation analysis in our laboratory was carried out by Lisa Prat Davies, computer molecular modelling and toxicity by Dr. Georgina Macintyre and Charles de Leeuw, MURR1 studies by Veronica Coronado, ATP7B yeast functional assays by Gloria Hsi, sequencing and database curation by Susan Kenney. Our copper studies are supported by CIHR (Canadian Institutes for Health Research) and NSERC (National Science and Engineering Research Council of Canada).

References

1. Wilson SAK. Progressive lenticular degeneration: a familial nervous disease associated with cirrhosis of the liver. Brain. 1912;34:295–509.
2. Yoshida K, Furihata K, Takeda S et al. A mutation in the ceruloplasmin gene is associated with systemic hemosiderosis in humans. Nat Genet. 1995;9:267–72.
3. Harris ZL, Takahashi Y, Miyajima H, Serizawa M, MacGillivray RT, Gitlin JD. Aceruloplasminemia: molecular characterization of this disorder of iron metabolism. Proc Natl Acad Sci USA. 1995;92:2539–43.
4. Field LS, Luk E, Culotta VC. Copper chaperones: personal escorts for metal ions. J Bioenerg Biomembr. 2002;34:373–9.

5. Sokol R, Twedt D, McKim JM et al. Oxidant injury to hepatic mitochondria in patients with Wilson's disease and Bedlington terriers with copper toxicosis. Gastroenterology. 1994;107:1788–98.
6. Ogihara H, Ogihara T, Miki M, Yasuda H, Mino M. Plasma copper and antioxidant status in Wilson's disease. Pediatr Res. 1995;37:219–26.
7. Roberts EA, Schilsky ML. A practice guideline on Wilson disease. Hepatology. 2003;37: 1475–92.
8. Cox DW, Roberts EA. Wilson disease. In: Scharschmidt B, Sleisinger MH, editors. Sleisinger and Fordtran's Gastrointestinal Disease: Pathophysiology, Diagnosis, Management, 8th edn. Philadelphia: WB Saunders; 2006:1269–77.
9. Cox DW, Roberts EA. Wilson disease. In: GeneReviews at GeneTests: Medical Genetics Information Resource. Seattle: University of Washington, 2006.
10. Roberts EA, Cox DW. Wilson disease. In: Boyer T, Manns M, Wright T, editors. Zakim and Boyer's Hepatology, 5th edn. Philadelphia: Elsevier; 2006.
11. Oder W, Prayer L, Grimm G. Wilson's disease: evidence of subgroups derived from clinical findings and brain lesions. Neurology. 1993;43:120–4.
12. Attri S, Sharma N, Jahagirdar S, Thapa BR, Prasad R. Erythrocyte metabolism and antioxidant status of patients with Wilson disease with hemolytic anemia. Pediatr Res. 2006;59:593–7.
13. Wilson DC, Phillips MJ, Cox DW, Roberts EA. Severe hepatic Wilson's disease in preschool-aged children. J Pediatr. 2000;137:719–22.
14. Ala A, Borjigin J, Rochwarger A, Schilsky M. Wilson disease in septuagenarian siblings: raising the bar for diagnosis. Hepatology. 2005;41:668–70.
15. Schilsky ML. Wilson disease: new insights into pathogenesis, diagnosis, and future therapy. Curr Gastroenterol Rep. 2005;7:26–31.
16. Sallie R, Katsiyiannakis L, Baldwin D. Failure of simple biochemical indexes to reliably differentiate fulminant Wilson's disease from other causes of fulminant liver failure. Hepatology. 1992;16:1206–11.
17. Cox DW. Factors influencing serum ceruloplasmin in normal individuals. J Lab Clin Med. 1966;68:893–904.
18. Steindl P, Ferenci P, Dienes HP et al. Wilson's disease in patients with liver disease: a diagnostic challenge. Gastroenterology. 1998;113:212–18.
19. Macintyre G, Gutfreund KS, Martin WR, Camicioli R, Cox DW. Value of an enzymatic assay for the determination of serum ceruloplasmin. J Lab Clin Med. 2004;144:294–301.
20. Walshe JM. Wilson's disease: the importance of measuring serum caeruloplasmin non-immunologically. Ann Clin Biochem. 2003;40:115–21.
21. Faa G, Nurchi V, Demelia L et al. Uneven hepatic copper distribution in Wilson's disease. J Hepatol. 1995;22:303–8.
22. Lyon TD, Fell GS, Gaffney D et al. Use of a stable copper isotope (^{65}Cu) in the differential diagnosis of Wilson's disease. Clin Sci. 1995;88:727–32.
23. Bull PC, Thomas GR, Rommens JM, Forbes JR, Cox DW. The Wilson disease gene is a putative copper transporting P-type ATPase similar to the Menkes gene. Nat Genet. 1993; 5:327–37.
24. Tanzi RE, Petrukhin KE, Chernov I et al. The Wilson disease gene is a copper transporting ATPase with homology to the Menkes disease gene. Nat Genet. 1993;5:344–50.
25. Thomas GR, Forbes JR, Roberts EA, Walshe JM, Cox DW. The Wilson disease gene: spectrum of mutations and their consequences [erratum in Nat Genet. 1995;9:451]. Nat Genet. 1995;9:210–17.
26. Hsi G, Cox DW. A comparison of the mutation spectra of Menkes disease and Wilson disease. Hum Genet. 2004;114:165–72.
27. Moller LB, Ott P, Lund C, Horn N. Homozygosity for a gross partial gene deletion of the C-terminal end of ATP7B in a Wilson patient with hepatic and no neurological manifestations. Am J Med Genet A. 2005;138:340–3.
28. Proud VK, Mussell HG, Kaler SG, Young DW, Percy AK. Distinctive Menkes disease variant with occipital horns: delineation of natural history and clinical phenotype. Am J Med Genet. 1996;65:44–51.
29. Kaler SG, Gallo LK, Proud VK et al. Occipital horn syndrome and a mild Menkes phenotype associated with splice site mutations at the MNK locus. Nat Genet. 1994;8: 195–202.

30. Caca K, Ferenci P, Kuhn HJ et al. High prevalence of the H1069Q mutation in East German patients with Wilson disease: rapid detection of mutations by limited sequencing and phenotype-genotype analysis. J Hepatol. 2001;35:575–81.
31. Loudianos G, Dessi V, Lovicu M et al. Molecular characterization of Wilson disease in the Sardinian population – evidence of a founder effect. Hum Mutat. 1999;14:294–303.
32. Cullen LM, Prat L, Cox DW. Genetic variation in the promoter and 5′ UTR of the copper transporter, ATP7B, in patients with Wilson disease. Clin Genet. 2003;64:429–32.
33. Huster D, Weizenegger M, Kress S, Mossner J, Caca K. Rapid detection of mutations in Wilson disease gene ATP7B by DNA strip technology. Clin Chem Lab Med. 2004;42:507–10.
34. Lovicu M, Dessi V, Zappu A, De Virgiliis S, Cao A, Loudianos G. Efficient strategy for molecular diagnosis of Wilson disease in the Sardinian population. Clin Chem. 2003;49: 496–8.
35. Panagiotakaki E, Tzetis M, Manolaki N et al. Genotype–phenotype correlations for a wide spectrum of mutations in the Wilson disease gene (ATP7B). Am J Med Genet. 2004;131A: 168–73.
36. Loudianos G, Lovicu M, Dessi V et al. Abnormal mRNA splicing resulting from consensus sequence splicing mutations of ATP7B. Hum Mutat. 2002;20:260–6.
37. Petris MJ, Voskoboinik I, Cater M et al. Copper-regulated trafficking of the Menkes disease copper ATPase is associated with formation of a phosphorylated catalytic intermediate. J Biol Chem. 2002;277:46736–42.
38. Schaefer M, Hopkins RG, Failla ML, Gitlin JD. Hepatocyte-specific localization and copper-dependent trafficking of the Wilson's disease protein in the liver. Am J Physiol. 1999;276:G639–46.
39. Roelofsen H, Wolters H, Van Luyn MJA, Miura N, Kuipers F, Vonk RJ. Copper-induced apical trafficking of ATP7B in polarized hepatoma cells provides a mechanism for biliary copper excretion. Gastroenterology. 2000;119:782–93.
40. Payne AS, Kelly EJ, Gitlin JD. Functional expression of the Wilson disease protein reveals mislocalization and impaired copper-dependent trafficking of the common H1069Q mutation. Proc Natl Acad Sci USA. 1998;95:10854–9.
41. Forbes JR, Cox DW. Functional characterization of missense mutations in ATP7B: Wilson disease mutation or normal variant? Am J Hum Genet. 1998;63:1663–74.
42. Payne AS, Gitlin JD. Functional expression of the Menkes disease protein reveals common biochemical mechanisms among the copper-transporting p-type ATPases. J Biol Chem. 1998;273:3765–70.
43. Forbes JR, Cox DW. Copper-dependent trafficking of Wilson disease mutant ATP7B proteins. Hum Mol Genet. 2000;9:1927–35.
44. Hsi G, Cullen LM, Moira Glerum D, Cox DW. Functional assessment of the carboxy-terminus of the Wilson disease copper-transporting ATPase, ATP7B. Genomics. 2004;83: 473–81.
45. Huster D, Hoppert M, Lutsenko S et al. Defective cellular localization of mutant ATP7B in Wilson's disease patients and hepatoma cell lines. Gastroenterology. 2003;124:335–45.
46. Cox DW, Prat L, Walshe JM, Heathcote J, Gaffney D. Twenty-four novel mutations in Wilson disease patients of predominantly European ancestry. Hum Mutat. 2005;26:280.
47. Fatemi N, Sarkar B. Structural and functional insights of Wilson disease copper-transporting ATPase. J Bioenerg Biomembr. 2002;34:339–49.
48. Lutsenko S, Efremov RG, Tsivkovskii R, Walker JM. Human copper-transporting ATPase ATP7B (the Wilson's disease protein): biochemical properties and regulation. J Bioenerg Biomembr. 2002;34:351–62.
49. Dmitriev O, Tsivkovskii R, Abildgaard F, Morgan CT, Markley JL, Lutsenko S. Solution structure of the N-domain of Wilson disease protein: distinct nucleotide-binding environment and effects of disease mutations. Proc Natl Acad Sci USA. 2006;103:5302–7.
50. Pieper U, Eswar N, Davis FP et al. MODBASE: a database of annotated comparative protein structure models and associated resources. Nucleic Acids Res. 2006;34:D291–5.
51. Gaffney D, Walker JL, O'Donnell JG et al. DNA-based presymptomatic diagnosis of Wilson disease. J Inherit Metab Dis. 1992;15:161–70.
52. Coronado VA, Bonneville JA, Nazer H, Roberts EA, Cox DW. COMMD1 (MURR1) as a candidate in patients with copper storage disease of undefined etiology. Clin Genet. 2005; 68:548–51.

53. Yuzbasiyan-Gurkan V, Blanton SH, Cao Y et al. Linkage of a microsatellite marker to the canine copper toxicosis locus in Bedlington terriers. Am J Vet Res. 1997;58:23–7.
54. van de Sluis B, Rothuizen J, Pearson PL, Van Oost BA, Wijmenga C. Identification of a new copper metabolism gene by positional cloning in a purebred dog population. Hum Mol Genet. 2002;11:165–73.
55. Coronado VA, Damaraju D, Kohijoki R, Cox DW. New haplotypes in the Bedlington terrier indicate complexity in copper toxicosis. Mammalian Genome. 2003;14:483–91.
56. Hyun C, Lavulo LT, Filippich LJ. Evaluation of haplotypes associated with copper toxicosis in Bedlington Terriers in Australia. Am J Vet Res. 2004;65:1573–9.
57. Tao TY, Liu F, Klomp L, Wijmenga C, Gitlin JD. The copper toxicosis gene product Murr1 directly interacts with the Wilson disease protein. J Biol Chem. 2003;278:41593–6.
58. Biasio W, Chang T, McIntosh CJ, McDonald FJ. Identification of Murr1 as a regulator of the human delta epithelial sodium channel. J Biol Chem. 2003;279:5429–34.
59. Burstein E, Ganesh L, Dick RD et al. A novel role for XIAP in copper homeostasis through regulation of MURR1. EMBO J. 2004;23:244–54.
60. Ganesh L, Burstein E, Guha-Niyogi A et al. The gene product Murr1 restricts HIV-1 replication in resting $CD4^+$ lymphocytes. Nature. 2003;426:853–7.
61. Muller T, van de Sluis B, Zhernakova A et al. The canine copper toxicosis gene MURR1 does not cause non-Wilsonian hepatic copper toxicosis. J Hepatol. 2003;38:164–8.
62. Stuehler B, Reichert J, Stremmel W, Schaefer M. Analysis of the human homologue of the canine copper toxicosis gene MURR1 in Wilson disease patients. J Mol Med. 2004;82:629–34.
63. Wu ZY, Zhao GX, Chen WJ et al. Mutation analysis of 218 Chinese patients with Wilson disease revealed no correlation between the canine copper toxicosis gene MURR1 and Wilson disease. J Mol Med. 2006;84:438–42.
64. Weiss KH, Merle U, Schaefer M, Ferenci P, Fullekrug J, Stremmel W. Copper toxicosis gene MURR1 is not changed in Wilson disease patients with normal blood ceruloplasmin levels. World J Gastroenterol. 2006;12:2239–42.
65. Lovicu M, Dessi V, Lepori MB et al. The canine copper toxicosis gene MURR1 is not implicated in the pathogenesis of Wilson disease. J Gastroenterol. 2006;41:582–7.

9
Wilson disease: therapeutic strategies

E. A. ROBERTS

Wilson disease is an inherited disorder of copper disposition which affects mainly the liver and brain, although other organs including the eyes, kidneys, erythrocytes, and endocrine organs are frequently involved. The pattern of inheritance is autosomal recessive. Clinical features of Wilson disease were first described in the early 1900s[1] and the pathophysiology of the disease elucidated over the next 40 years. The gene which is mutated in Wilson disease was identified in 1993[2–4]. This gene, *ATP7B*, encodes a metal-transporting P-type ATPase (called the Wilson ATPase) which spans intracellular membranes and plays a key role in the intracellular disposition of copper. To date nearly 300 mutations have been identified. Most patients are compound heterozygotes who bear one each of two different mutations. Ongoing research indicates that Wilson disease is a single-gene disorder. Modifier genes have not been extensively identified. Genotype–phenotype correlations are imperfect. The basis for the clinical variability remains unexplained.

Wilson disease is considered to be rather rare. It occurs worldwide. On average there are 30 cases/million population, but some geographically isolated communities have higher prevalence. Diagnosis is often difficult because of the highly variable clinical presentations of Wilson disease; thus its rarity may be overestimated. Wilson disease can present as a hepatic, neurological or psychiatric disorder. Although the classic age range for development of clinically apparent disease is 6–45-years-old, pre-school-aged children can develop serious liver disease[5,6], and Wilson disease has been diagnosed in adults in their 60s and 70s[7]. In children the disease presentation may be especially subtle[8,9]. Hepatic disease tends to be more prominent in younger patients whereas neurological or neuropsychiatric disease tends to be more frequent with older patients. In women of child-bearing age, unexplained infertility or repeated miscarriages may be due to undiagnosed, and thus untreated, Wilson disease. Mild-to-moderate haemolytic anaemia may also be the first clinical evidence of Wilson disease. Diagnosis is easy if the patient is in the typical age bracket, has classic hepatic or neurological disease, has well-formed Kayser–Fleischer rings on slit-lamp examination, and a very low (<50 mg/L) serum concentration of caeruloplasmin. Such patients are rather uncommon. Extensive clinical investigation, including measurement of basal 24-h urinary copper excretion, cerebral imaging, and liver biopsy for histological examination and quantification of parenchymal copper concentration, is usually necessary[10]. Liver disorders associated with Wilson disease range from minor liver test abnormalities to fatty liver to a clinical mimic of autoimmune hepatitis[11] to cirrhosis. This means that Wilson disease must be borne in mind for any patient with unexplained liver disease, and certainly for all children with liver disease. A few patients present with Wilsonian fulminant hepatic failure, a classic variant of acute liver failure[12,13].

Ideally, treatment should be based on disease pathophysiology. In Wilson disease copper accumulates in the liver because it is not handled properly in hepatocytes. The Wilson ATPase plays a role in making copper available for incorporation into caeruloplasmin[14] and, when hepatocellular concentrations of copper are elevated, expediting the biliary excretion of copper[15]. Depending on the mutation, one or both of these functions may be compromised. Copper can participate in Fenton-like reactions and thus copper overload is associated with oxidative stress[16]. Mitochondrial damage is prominent in Wilson disease. Copper-overloaded hepatocytes may be predisposed to apoptosis[17,18]. Thus the direct treatment strategy is to remove copper from the liver and elsewhere in the body, and the indirect strategies include providing antioxidants, and possibly interfering with hepatic apoptosis.

Therapeutics of Wilson disease is exceptional by contemporary standards because the drugs are mostly orphan drugs, and they have not been subjected to extensive systematic clinical testing. There is a surprising paucity of well-designed clinical pharmacological research, and treatment protocols often rest on cumulative experience with comparatively small patient numbers. Nevertheless drug treatment of Wilson disease is generally effective, and the risks and benefits of the drugs used are constantly being re-evaluated. Currently available treatments for Wilson disease include orally administered chelators: D-pencillamine and trientine. The other generally accepted treatment is zinc, which extensively inhibits copper absorption in the intestine and also induces metallothioneins in the gut and liver. Ammonium tetrathiomolybdate, used only as an investigational drug, both inhibits copper absorption and complexes copper within the plasma compartment. This can lead to profound copper depletion. Recently, various combination regimens have been tried clinically. Vitamin E is sometimes used as an antioxidant, but there are no rigorous studies as to its effectiveness. The utility of complementary medicine, especially naturally occurring flavonoids, has not been investigated. Liver transplantation is reserved for the most severe cases of hepatic Wilson disease. A low-copper diet alone is not adequate treatment; however, excessively high concentrations of copper in drinking water should be avoided. People with Wilson disease should not drink alcohol since it may itself cause mitochondrial damage. Maintaining a healthy body weight is probably beneficial, although there are no specific data on this point.

Making the accurate diagnosis is the first step to treating Wilson disease. Treatment is lifelong and involves some risk of adverse side-effects. Accordingly, it is critically important not to consign a patient to such treatment inappropriately. Equally important, untreated Wilson disease has a grave prognosis. Commencing treatment as early as possible provides the best outcome[19]. Once the diagnosis is made, the severity of liver disease is an important consideration for choosing treatment. Mild-to-moderate liver disease may be treated with a copper-depleting agent (chelator or possibly zinc) alone. Decompensated cirrhosis is increasingly treated with combination therapy, usually trientine and zinc, given in a temporally dispersed regimen. Each agent is given separately in an alternating fashion through the 24-h day: in effect, one or the other is administered every 5–6 h. This intensive regimen is generally used for 3–6 months until the hepatic disease is stabilized. After this, the patient is treated with a single-agent standard regimen. Wilsonian fulminant hepatic failure requires liver transplantation urgently. Albumin dialysis may be used as a bridge to transplant. The degree of neurological involvement also has to be factored into choice of regimen. Patients with mainly neurological symptoms may worsen when started on D-penicillamine. Primary treatment with trientine or zinc is advocated. The safety and effectiveness of tetrathiomolybdate combined with zinc is being studied as initial treatment for neurological Wilson disease. Patients who already have diseases affecting the kidneys, bone marrow, or skin, or who have an autoimmune diathesis, should not be treated with D-penicillamine, since these organs most often are the targets of its most severe adverse effects. In addition to being used with patients who present with fulminant hepatic failure, liver transplantation is reserved for those who do not respond to medical therapy (a relatively small proportion) or those who become non-compliant with the treatment regimen and develop chronic liver failure.

Penicillamine has been the first-line treatment for Wilson disease for decades since it was first used by Walshe in the mid-1950s[20]. It is effective in most patients, and it has been life-saving for many patients. Penicillamine is β,β-dimethylcysteine; that is, the amino acid cysteine, which has a sulphydryl group, substituted with two methyl groups. It can be taken by mouth, and it is absorbed rapidly with moderately good bioavailability on the order of 50%. Penicillamine depletes total body copper by greatly increasing urinary excretion of copper. Penicillamine is a reductive chelator. It reduces protein-bound Cu(II): the resulting Cu(I) has less affinity for the protein and can then be chelated to penicillamine[21]. Penicillamine can remove copper from tissues but it does not remove all copper from the liver, probably because it mainly targets lysosomal copper in hepatocytes. Penicillamine has some other pharmacological actions: it inhibits collagen crosslinking (a problem with chronic administration in Wilson disease) and it is an immunosuppressant. Penicillamine has been used to treat various rheumatological disorders; it is ineffective in chronic cholestatic liver disease; it is a standard treatment for cystinuria.

As a drug treatment penicillamine is limited by having serious adverse side-effects. Overall, D-penicillamine is less toxic than the racemic mixture and is the form of penicillamine currently used. An immediate adverse reaction is a febrile reaction with rash and proteinuria which some patients develop within the first week or so after beginning treatment. Although D-penicillamine can be restarted slowly, along with corticosteroids, changing to an alternative chelator may be safer and has become the customary management. A serious early adverse effect is that the neurological status of patients presenting with neurological Wilson disease may deteriorate severely soon after D-penicillamine treatment is started. This occurs in 20–40%, or more[22], of these patients. Most, but not all, patients recover entirely or regain most function even with continued use of D-penicillamine. Of interest, structural changes in the central nervous system have been reported in patients who deteriorate[23]. Using D-penicillamine as the first-line drug for patients with neurological Wilson disease is probably not appropriate. If it is used, and neurological deterioration occurs, the current practice is to discontinue the D-penicillamine and substitute an alternate treatment, usually zinc[24].

Important adverse effects can also develop in the medium or long term, even in patients who appear entirely stable on D-penicillamine treatment. These include proteinuria, leucopenia, and thrombocytopenia. Aplastic anaemia occurs rarely and does not always reverse when D-penicillamine is stopped. Nephrotic syndrome, Goodpasture syndrome, and a systemic disease resembling lupus erythematosus have all been reported. A condition resembling myasthenia gravis can develop; serous retinitis and hepatotoxicity have been reported. All these severe side-effects require immediate discontinuation of D-penicillamine and use of a different treatment, usually trientine. Approximately one-third of Wilson disease patients develop a severe side-effect of D-penicillamine that mandates change of drug[25]. Regular clinical review of all patients on D-penicillamine is essential. Adverse reactions involving the skin include various rashes, pemphigus and related pemphigoid disorders, elastosis perforans serpiginosa and cutis laxa. Chronic administration of D-penicillamine can deplete pyridoxine stores and cause other toxicities such optic neuritis; pyridoxine supplementation is routine with D-penicillamine use. Relatively minor side-effects include loss of taste, gastrointestinal upset and arthralgias.

Although long-term toxicity is attributed to the D-penicillamine, it is really not clear whether all these effects represent intrinsic drug toxicity or some subtle effect of changes in copper disposition and availability to tissues. With 50 years' experience in using this drug, it is time to assess the safety of lifelong treatment especially for the skin and connective tissues. Chronic depletion of copper may also occur, and it usually results in neurological problems or anaemia.

Trientine (triethylene tetramine dihydrochloride; 2,2,2-tetramine; official short chemical name: trien) is a chelator chemically distinct from D-penicillamine; it was introduced by Walshe in the early 1980s[26]. It remains the usual second-line treatment for patients who cannot tolerate D-penicillamine. Trientine is chemically very different from D-penicillamine. It is a polyamine chelator, with an entirely different structure and no sulphydryl groups. Copper is chelated because it forms a stable complex with the four constituent nitrogens in a planar ring. The action of trientine is mainly to increase urinary copper excretion, but it may also interfere with intestinal absorption of copper[27]. It targets copper in the plasma compartment. Although trientine is a somewhat less potent chelator than D-penicillamine, the difference is not important clinically. Like D-penicillamine it is a relatively non-specific chelator.

Trientine has few adverse side-effects, apart from causing occasional gastritis and inducing iron deficiency, apparently by chelating dietary iron. Severe bone marrow suppression has been reported rarely. Sideroblastic anaemia has been reported, but appears to be due to copper deficiency, not the drug itself. Neurological worsening after beginning treatment with trientine has rarely been reported. Problems associated with D-penicillamine resolve when patients are changed over to trientine treatment[28]. There is growing enthusiasm for using trientine as an effective first-line treatment, especially in those patients with pre-existing bone marrow or kidney disease. Trientine is well tolerated[29,30].

Zinc has been used mainly as zinc sulphate in Europe since the late 1960s[31] and as zinc acetate in North America for the past 20 years. Zinc gluconate has been used in Canada for approximately 15 years. The actual salt appears to affect tolerability more than efficacy, although purity of zinc preparation is always important. Given in large doses, zinc interferes with absorption of copper from the gastrointestinal tract and increases copper excretion in the stools. Although it competes with copper for uptake into enterocytes, its main action is to induce metallothionein in enterocytes. This metallothionein has a greater affinity for copper than for zinc and preferentially binds copper from the intestinal contents. Once bound, the copper is not absorbed and is excreted in the faeces as enterocytes are shed in normal turnover[32]. Copper in the intestinal contents includes non-caeruloplasmin bound copper which has been secreted into gastric juice, pancreatic juice and bile; thus chronic zinc treatment causes a net loss of copper from the body. Zinc may also induce hepatocellular metallothionein and thus stabilize hepatocellular copper; hepatic parenchymal copper concentration does not decrease during treatment with zinc[33].

Zinc treatment seems effective and relatively non-toxic[34–37]. In a randomized controlled trial comparing D-penicillamine and zinc sulphate, both were equally effective in stabilizing the clinical disease, but adverse side-effects were more numerous with D-penicillamine[38]. However, two patients with hepatic Wilson disease deteriorated when zinc was used as primary therapy. Gastritis is the most common side-effect, and this is bothersome to occasional patients. Food interferes with the effectiveness of zinc. If it is taken well away from meals the gastritis may be worse. Any cumbersome treatment regimen risks compromising compliance in adolescents and young adults. Treatment should be individualized to obtain greatest efficacy; this may involve increasing the dose.

Rare patients have experienced significant deterioration in hepatic Wilson disease when started on zinc. Studies in rats suggest possible interference with bone formation. Immune function may be depressed. The long-term effects of zinc excess are unclear and may disturb lipid metabolism. In general these toxicities appear fairly minor[39].

Combining zinc with a conventional chelator, either D-penicillamine or trientine, has recently become a popular treatment strategy for especially severe hepatic Wilson disease; that is, decompensated cirrhosis at presentation, usually indicated by evidence of severely compromised hepatic synthetic function (hypoalbuminaemia, prominent coagulopathy), ascites, but no encephalopathy. In order to avoid having chelator bind the zinc, the two types of treatments must be temporally dispersed through the day, with usually 5–6 h between administration of either drug. Our routine is to alternate the drugs through the day at approximately 6-hourly intervals: zinc 50 mg elemental is given as the first and third doses, and trientine (500 mg, or ~10 mg/kg in children) given as the second and fourth doses. This intensive treatment may thus be best suited to very severe hepatic or neurological disease; indeed this is where it has mainly been used[40,41]. Clinical data are limited but seem to favour the combination of trientine with zinc; nevertheless, this strategy should be regarded as investigational. The patient with severe hepatic decompensation but not all the features of Wilsonian fulminant hepatic failure may benefit from a trial of this regimen (Roberts EA, unpublished observations)[5,42], although some may not respond to this medical treatment and still require liver transplantation. This is fundamentally an induction regimen, and once the liver disease has improved, usually after approximately 3–6 months of intensive treatment, the patient is placed on a standard single-agent regimen: our routine is to use trientine alone.

Ammonium tetrathiomolybdate is an investigational treatment for Wilson disease and is reserved for neurological Wilson disease. It is not associated with early neurological deterioration[43]. It is a very potent agent for removing copper. Tetrathiomolybdate both interferes with copper absorption from the intestine and chelates copper loosely bound to plasma proteins to form a stable complex composed of the copper, tetrathiomolybdate, and albumin so that copper is not available for uptake into cells. It is really not known whether tetrathiomolybdate is non-toxic. Bone marrow suppression is an important and potentially serious adverse side-effect, possibly due to copper deficiency[44]. Serum aminotransferase elevations may occur with treatment, occasionally severe[45]. Dose and length of treatment, as well as long-term side-effects, must still be determined, but short-term use in combination with zinc for neurological presentation may be the optimal role for this drug[46]. Systemic copper deficiency might develop with prolonged use of this potent copper-binding drug, but this is being investigated as a strategy to suppress tumour angiogenesis[47]. Cerebral deposition of molybdate is associated with an organic brain syndrome.

Vitamin E (α-tocopherol) supplementation may be useful in Wilson disease, but no rigorous data are available. Patients have been reported to have low serum levels of vitamin E in the presence of high concentrations of non-caeruloplasmin-bound copper[48,49].

Liver transplantation plays a limited but important role in Wilson disease. Fulminant hepatic failure in Wilson disease always necessitates liver transplantation for recovery. Some patients with severe liver disease that is unresponsive to drug therapy may also proceed to transplantation. Using liver transplantation to treat severe neurological disease is a contentious issue[50–52]. Living-related transplants, where the graft may be from a heterozygote, have been found to function adequately despite minor defects in copper disposition sometimes encountered in heterozygotes[53,54]. Survival after liver transplantation for Wilson disease is highly favourable, on the order of 70–88% or more[55]. Patients with Wilsonian fulminant hepatic failure may benefit from procedures such as haemofiltration or albumin dialysis[56] to protect against progressive renal damage and extend survival until a suitable liver graft can be found.

Hepatocyte transplantation and gene therapy may be therapeutic interventions in the future, but further technical development is required before this is a practical option[57–59].

In pregnancy, patient surveillance should be intensified, but treatment should not be discontinued because of the risk of postpartum hepatic decompensation if treatment is stopped altogether. Although D-penicillamine has been used during numerous pregnancies in women with Wilson disease, and is considered safe[60], it is classified as having a 5% risk of fetotoxicity. Severe collagen defects in the baby have occasionally been reported[61]. Trientine and zinc have also been used during pregnancy, but data are relatively limited[62,63]. If either D-penicillamine or trientine is used, the dose should be decreased, by approximately 25% of the pre-pregnancy dose, in the third trimester, especially if Caesarean section is likely. Adequate diet, iron sufficiency and folic acid supplementation are very important in the pregnant Wilson disease patient. Depending on the severity of liver disease, delivery in a high-risk maternity unit may be appropriate. Women taking D-penicillamine should not breast-feed since the drug gets into breast milk and might harm the infant. Little is known about the safety of trientine and zinc in breast milk.

Once commenced, treatment with some drug to control copper accumulation in Wilson disease is mandatory throughout life. If intolerance develops to one drug, another must be substituted. Prolonged non-compliance with drug treatment leads to irreversible liver failure. Nonetheless the long-term risks associated with any treatment for Wilson disease remain unclear. Apart from the skin changes characteristic of penicillamine treatment, deficiencies in trace metals may develop if chelation is relatively non-specific. If caeruloplasmin oxidase activity remains very low for decades, abnormal iron metabolism causing hepatic iron overload and anaemia may develop[64].

With the advent of oral chelators in the 1950s and 1960s, Wilson disease was no longer a lethal condition. Nowadays patients with Wilson disease who are diagnosed promptly and treated effectively have a good prognosis. Severe neurological disease may not resolve entirely on treatment or with liver transplantation. The presymptomatic individual diagnosed on biochemical or genetic grounds has the best long-term prognosis with early treatment. Thus screening the family is vitally important. General screening of all newborns poses special problems, however, since copper depletion during the critical early growth period may be deleterious to overall growth and development. Our growing knowledge of the cellular physiology of copper handling may lead to new therapeutic modalities. What distinguishes the treatment of Wilson now compared to even 25 years ago is that we now have a broad armamentarium of drugs plus the availability of liver transplantation: therefore, treatment can be customized to the needs of the individual patient. Further specifics regarding treatment of Wilson disease are found in a recent practice guideline[10].

References

1. Wilson SAK. Progressive lenticular degeneration: a familial nervous disease associated with cirrhosis of the liver. Brain. 1912;34:295–507.
2. Bull PC, Thomas GR, Rommens JM et al. The Wilson disease gene is a putative copper transporting P-type ATPase similar to the Menkes gene. Nat Genet. 1993;5:327–37.
3. Tanzi RE, Petrukhin K, Chernov I et al. The Wilson disease gene is a copper transporting ATPase with homology to the Menkes disease gene. Nat Genet. 1993;5:344–50.
4. Petrukhin K, Fischer SG, Pirastu M et al. Mapping, cloning and genetic characterization of the region containing the Wilson disease gene. Nat Genet. 1993;5:338–43.
5. Wilson DC, Phillips MJ, Cox DW et al. Severe hepatic Wilson's disease in preschool-aged children. J Pediatr. 2000;137:719–22.
6. Beyersdorff A, Findeisen A. Morbus Wilson: case report of a two-year-old child as first manifestation. Scand J Gastroenterol. 2006;41:496–7.
7. Ala A, Borjigin J, Rochwarger A et al. Wilson disease in septuagenarian siblings: raising the bar for diagnosis. Hepatology. 2005;41:668–70.
8. Lo Curto AG, Marchi A, Grasso M et al. Early diagnosis of Wilson disease in a six-year-old child. J Pediatr. 2006;148:141.
9. Caprai S, Loudianos G, Massei F et al. Direct diagnosis of Wilson disease by molecular genetics. J Pediatr. 2006;148:138–40.
10. Roberts EA, Schilsky ML. A practice guideline on Wilson disease. Hepatology. 2003;37:1475–92.
11. Milkiewicz P, Saksena S, Hubscher SG et al. Wilson's disease with superimposed autoimmune features: report of two cases and review. J Gastroenterol Hepatol. 2000;15:570–4.
12. Roche-Sicot J, Benhamou JP. Acute intravascular hemolysis and acute liver failure associated as a first manifestation of Wilson's disease. Ann Intern Med. 1977;86:301–3.
13. Herrmann S, Hofmann W, Theilmann L. [Acute liver failure as the initial manifestation of Wilson disease]. Med Klin. 1995;90:456–61.
14. Tao TY, Gitlin JD. Hepatic copper metabolism: insights from genetic disease. Hepatology. 2003;37:1241–7.
15. Cater MA, La Fontaine S, Shield K et al. ATP7B mediates vesicular sequestration of copper: insight into biliary copper excretion. Gastroenterology. 2006;130:493–506.
16. Sokol RJ, Twedt D, McKim JM Jr et al. Oxidant injury to hepatic mitochondria in patients with Wilson's disease and Bedlington terriers with copper toxicosis. Gastroenterology. 1994;107:1788–98.
17. Deng DX, Ono S, Koropatnick J et al. Metallothionein and apoptosis in the toxic milk mutant mouse. Lab Invest. 1998;78:175–83.
18. Strand S, Hofmann WJ, Grambihler A et al. Hepatic failure and liver cell damage in acute Wilson's disease involve CD95 (APO-1/Fas) mediated apoptosis. Nat Med. 1998;4:588–93.
19. Sternlieb I, Scheinberg IH. Prevention of Wilson's disease in asymptomatic patients. N Engl J Med. 1968;278:352–9.
20. Walshe JM. The story of penicillamine: a difficult birth. Mov Disord. 2003;18:853–9.
21. Sarkar B. Treatment of Wilson and Menkes diseases. Chem Rev. 1999;99:2535–44.
22. Brewer GJ, Terry CA, Aisen AM et al. Worsening of neurologic syndrome in patients with Wilson's disease with initial penicillamine therapy. Arch Neurol. 1987;44:490–3.
23. Huang CC, Chu NS. Acute dystonia with thalamic and brainstem lesions after initial penicillamine treatment in Wilson's disease. Eur Neurol. 1998;39:32–7.
24. Huang CC, Chu NS. Wilson's disease: resolution of MRI lesions following long-term oral zinc therapy. Acta Neurol Scand. 1996;93:215–18.
25. Walshe JM. Wilson's disease presenting with features of hepatic dysfunction: a clinical analysis of eighty-seven patients. Q J Med. 1989;70:253–63.
26. Walshe JM. Treatment of Wilson's disease with trientine (triethylene tetramine) dihydrochloride. Lancet. 1982;1:643–7.
27. Siegemund R, Lossner J, Gunther K et al. Mode of action of triethylenetetramine dihydrochloride on copper metabolism in Wilson's disease. Acta Neurol Scand. 1991;83:364–6.
28. Scheinberg IH, Jaffe ME, Sternlieb I. The use of trientine in preventing the effects of interrupting penicillamine therapy in Wilson's disease. N Engl J Med. 1987;317:209–13.

29. Dahlman T, Hartvig P, Lofholm M et al. Long-term treatment of Wilson's disease with triethylene tetramine dihydrochloride (trientine). Q J Med. 1995;88:609-16.
30. Dubois RS, Rodgerson DO, Hambidge KM. Treatment of Wilson's disease with triethylene tetramine hydrochloride (trientine). J Pediatr Gastroenterol Nutr. 1990;10:77–81.
31. Hoogenraad TU, Van Hattum J, Van den Hamer CJ. Management of Wilson's disease with zinc sulphate. Experience in a series of 27 patients. J Neurol Sci. 1987;77:137–46.
32. Yuzbasiyan-Gurkan V, Grider A, Nostrant T et al. Treatment of Wilson's disease with zinc: X. Intestinal metallothionein induction. J Lab Clin Med. 1992;120:380–6.
33. Lee DY, Brewer GJ, Wang YX. Treatment of Wilson's disease with zinc. VII. Protection of the liver from copper toxicity by zinc-induced metallothionein in a rat model. J Lab Clin Med. 1989;114:639–45.
34. Brewer GJ, Dick RD, Johnson VD et al. Treatment of Wilson's disease with zinc: XV long-term follow-up studies. J Lab Clin Med. 1998;132:264–78.
35. Brewer GJ, Dick RD, Johnson VD et al. Treatment of Wilson's disease with zinc XVI: treatment during the pediatric years. J Lab Clin Med. 2001;137:191–8.
36. Anderson LA, Hakojarvi SL, Boudreaux SK. Zinc acetate treatment in Wilson's disease. Ann Pharmacother. 1998;32:78–87.
37. Hoogenraad TU. Paradigm shift in treatment of Wilson's disease: zinc therapy now treatment of choice. Brain Dev. 2006;28:141–6.
38. Czlonkowska A, Gajda J, Rodo M. Effects of long-term treatment in Wilson's disease with D-penicillamine and zinc sulphate. J Neurol. 1996;243:269–73.
39. Brewer GJ, Yuzbasiyan-Gurkan V, Johnson V. Treatment of Wilson's disease with zinc. IX: response of serum lipids. J Lab Clin Med. 1991;118:466–70.
40. Askari FK, Greenson J, Dick RD et al. Treatment of Wilson's disease with zinc. XVIII. Initial treatment of the hepatic decompensation presentation with trientine and zinc. J Lab Clin Med. 2003;142:385–90.
41. Santos Silva EE, Sarles J, Buts JP et al. Successful medical treatment of severely decompensated Wilson disease. J Pediatr. 1996;128:285–7.
42. Durand F, Bernuau J, Giostra E et al. Wilson's disease with severe hepatic insufficiency: beneficial effects of early administration of D-penicillamine. Gut. 2001;48:849–52.
43. Brewer GJ, Hedera P, Kluin KJ et al. Treatment of Wilson disease with ammonium tetrathiomolybdate: III. Initial therapy in a total of 55 neurologically affected patients and follow-up with zinc therapy. Arch Neurol. 2003;60:379–85.
44. Karunajeewa H, Wall A, Metz J et al. Cytopenias secondary to copper depletion complicating ammonium tetrathiomolybdate therapy for Wilson's disease. Aust NZ J Med. 1998;28:215–16.
45. Medici V, Trevisan CP, Bigotto MA et al. Adverse reaction after tetrathiomolybdate treatment for Wilson's disease: a case report. Mov Disord. 2006;21:2030–2.
46. Brewer GJ, Askari F, Lorincz MT et al. Treatment of Wilson disease with ammonium tetrathiomolybdate: IV. Comparison of tetrathiomolybdate and trientine in a double-blind study of treatment of the neurologic presentation of Wilson disease. Arch Neurol. 2006;63: 521–7.
47. Brewer GJ. Tetrathiomolybdate anticopper therapy for Wilson's disease inhibits angiogenesis, fibrosis and inflammation. J Cell Mol Med. 2003;7:11–20.
48. von Herbay A, de Groot H, Hegi U et al. Low vitamin E content in plasma of patients with alcoholic liver disease, hemochromatosis and Wilson's disease. J Hepatol. 1994;20:41–6.
49. Sinha S, Christopher R, Arunodaya GR et al. Is low serum tocopherol in Wilson's disease a significant symptom? J Neurol Sci. 2005;228:121–3.
50. Stracciari A, Tempestini A, Borghi A et al. Effect of liver transplantation on neurological manifestations in Wilson disease. Arch Neurol. 2000;57:384–6.
51. Medici V, Mirante VG, Fassati LR et al. Liver transplantation for Wilson's disease: the burden of neurological and psychiatric disorders. Liver Transpl. 2005;11:1056–63.
52. Senzolo M, Loreno M, Fagiuoli S et al. Different neurological outcome of liver transplantation for Wilson's disease in two homozygotic twins. Clin Neurol Neurosurg. 2007;109:71–5.
53. Komatsu H, Fujisawa T, Inui A et al. Hepatic copper concentration in children undergoing living related liver transplantation due to Wilsonian fulminant hepatic failure. Clin Transplant. 2002;16:227–32.

54. Wang XH, Cheng F, Zhang F et al. Living-related liver transplantation for Wilson's disease. Transplant Int. 2005;18:651–6.
55. Sutcliffe RP, Maguire DD, Muiesan P et al. Liver transplantation for Wilson's disease: long-term results and quality-of-life assessment. Transplantation. 2003;75:1003–6.
56. Kreymann B, Seige M, Schweigart U et al. Albumin dialysis: effective removal of copper in a patient with fulminant Wilson disease and successful bridging to liver transplantation: a new possibility for the elimination of protein-bound toxins. J Hepatol. 1999;31:1080–5.
57. Allen KJ, Cheah DM, Wright PF et al. Liver cell transplantation leads to repopulation and functional correction in a mouse model of Wilson's disease. J Gastroenterol Hepatol. 2004; 19:1283–90.
58. Park SM, Vo K, Lallier M et al. Hepatocyte transplantation in the Long Evans Cinnamon rat model of Wilson's disease. Cell Transplant. 2006;15:13–22.
59. Misawa R, Ise H, Takahashi M et al. Development of liver regenerative therapy using glycoside-modified bone marrow cells. Biochem Biophys Res Commun. 2006;342:434–40.
60. Sternlieb I. Wilson's disease and pregnancy. Hepatology. 2000;31:531–2.
61. Pinter R, Hogge WA, McPherson E. Infant with severe penicillamine embryopathy born to a woman with Wilson disease. Am J Med Genet A. 2004;128:294–8.
62. Walshe JM. The management of pregnancy in Wilson's disease treated with trientine. Q J Med. 1986;58:81–7.
63. Brewer GJ, Johnson VD, Dick RD et al. Treatment of Wilson's disease with zinc. XVII: treatment during pregnancy. Hepatology. 2000;31:364–70.
64. Shiono Y, Wakusawa S, Hayashi H et al. Iron accumulation in the liver of male patients with Wilson disease. Am J Gastroenterol. 2001;96:3147–51.

10
Alpha-1-antitrypsin deficiency: a journey through 30 years, clinical findings in relation to current biochemical knowledge

T. SVEGER

INTRODUCTION

The discovery of alpha-1-antitrypsin (AT) deficiency in 1962 by Laurell, followed by Sten Eriksson's thesis in 1965 outlining the association with lung emphysema, started a spectacular development in this field of inborn errors of metabolism. Harvey Sharp and co-workers in 1969 described the connection of AT deficiency with liver cirrhosis. Today AT deficiency is recognized as a member of the serpinopathies and conformational diseases. Excellent reviews have recently been published[1–4].

The prognosis into young middle age of the PiZ AT-deficient individuals, identified in the Swedish neonatal screening, has been surprisingly positive. They have normal lung function and none has clinical symptoms of liver disease[5] (also E. Bernspång, personal communication). Analyses of protease inhibitors and neutrophil lipocalin, a marker of neutrophil activity, have indicated compensatory changes in the AT deficiency state[6]. The great majority of AT-deficient subjects, most of them lifelong non-smokers, will live a normal life without clinical signs of emphysema[7]. Thus in a non-smoking PiZ generation liver disease in childhood and late adulthood will be the major clinical problems.

SELECTIVE HETEROZYGOTE ADVANTAGE

When we started the neonatal AT screening study, more than 30 years ago, the main purposes were to prevent lung disease and try to understand the natural history of liver disease and its clinical variation. Another question that puzzled me was why do we have such a high prevalence of AT deficiency; what is the advantage of being one of the 5% Scandinavians heterozygous for the PiZ gene[8]? In northern Italy and Curland an even higher MZ heterozygosity has been found[9]. Over the years we eventually found one advantage in our psychological studies, while David Lomas has recently presented a more sophisticated hypothesis. When families with an AT-deficient child were compared with carefully matched control families, the AT-deficient families had significantly higher socioeconomic class levels[10]. It therefore seems likely that the Vikings not only brought with them the PiZ gene, but that this gene may be coupled with a personality trait positively affecting survival and reproduction[11].

David Lomas focuses on the fact that the mutant S and Z AT form polymers within alveoli and small airways where they may cause inflammation and subsequently emphysema[12]. In the preantibiotic era this may have been an advantage amplifying the inflammatory response and limiting invasive respiratory and gastrointestinal infections. Smoking and increased longevity have turned an ancient protective advantage into today's detrimental effects, specifically emphysema[13,14].

INTRAUTERINE LIFE

At birth 45% of the 22 PiZ infants in the Swedish screening study with clinical signs of liver disease were small for gestational age, SGA[15]. The corresponding percentages for clinically healthy PiZ and PiSZ infants were 17% and 4%, respectively.

Fetal growth is dependent upon mother, placenta and fetus[16]. The growth retardation of the PiZ infants is most probably caused by the fetus itself. Liver cell inclusion bodies of polymerized Z protein occur already in intrauterine life[17]. In 14–24-week-old fetuses the AT concentration is normal, antichymotrypsin occurs in trace amounts, while alpha-2-macroglobulin starts at 30% and then increases up to 150% of the normal adult concentration at delivery in term infants[18]. Since AT Z protein is synthesized in the fetal liver it is logical that the mutant Z protein aggregates in the endoplasmic reticulum to form large polymers[19]. Only 15% of the Z protein is secreted while the main part is recognized by the quality control system within the endoplasmic reticulum and directed for proteolysis[20]. It is the accumulated Z protein within the hepatocytes that causes liver disease[21]. We do not know why four to five times more PiZ infants are growth-retarded compared with infants with normal AT, but a plausible explanation is liver cell injury by Z protein and the subsequent responses recently reviewed by Teckman and Lindblad[22].

INFANCY

The Swedish neonatal screening study has given us an excellent survey of the natural history of liver abnormalities in early childhood[15,23]. Of the 120 PiZ infants nine had both clinical and biochemical signs of neonatal cholestasis, while five had only bilirubin and liver enzyme abnormalities. Another eight infants had other clinical symptoms without jaundice. Of the 98 healthy PiZ infants 52% and 50% respectively had abnormal liver tests at 3 and 6 months of life.

Even in infants with normal AT, the first months of life is a period with a tendency towards cholestasis[24]. Infants with AT deficiency may be particularly vulnerable. Unusual hydroxylated bile acids occur both in meconium and during the first weeks of life[25]. Certain bile acids may be harmful in PiZ infants by increasing the intracellular Z protein polymerization[26].

The combination of intrauterine growth retardation and subsequent malabsorption due to cholestasis is detrimental to the PiZ infants. Teckman and co-workers examined the effect of fasting in a PiZ mouse model[27]. Fasting induced both an increased accumulation of fat in the liver and of AT-containing globules. Remarkably, PiZ mice had a markedly decreased fasting tolerance; 0% survival after 72 h compared with 100% in wild-type mice.

The composition, not only the energy content, of nutrients is important. An early study by Udall et al. suggested that breast milk may prevent severe liver disease in AT-deficient infants[28]. Feeding histories were also available from 118 of the Swedish PiZ infants. Of the 47 infants breast-fed less than a month 26% had clinical symptoms of liver disease compared with 11% in infants fed breast milk more than a month[29]. Serum γ-glutamyl transferase was also lower at 2 and 4 months of life in breast-fed infants. The way in which bioactive components of breast milk influence and regulate the immune system and inflammatory response in AT-deficient infants has not been explored[30]. The high risk for vitamin K deficiency and severe bleeding diathesis must never be forgotten.

CHILDHOOD

Two of the children with severe neonatal cholestasis, both small for gestational age, deteriorated after a short period of improvement. At 2–4 years of age they had severe symptoms of liver cirrhosis. They both died 7 years old[23,31,32]. A third PiZ child, also having suffered from neonatal cholestasis, healthy thereafter, died of aplastic anaemia, the autopsy revealing liver cirrhosis. A fourth PiZ child also having had neonatal cholestasis died in a traffic accident. Liver microscopy showed only a mild increase in periportal liver tissue. At 2, 4 and 8 years about 40–50% of the clinically healthy PiZ children had abnormal S-ALAT levels. At 8 years of age 6% had marginally increased S-GT concentrations. Notably the PiZ and SZ children had transferase levels within the normal range.

The PiZ children up to age 8 years demonstrate a remarkable variation in clinical and biochemical signs of liver disease. Only 2.5% died of or with liver cirrhosis, about one-fourth of the neonatally cholestatic infants, while the great majority remained clinically healthy, often with abnormal liver test results. Recent extensive reviews give the biochemical background of our clinical observations[19,22,33]. The familiar occurrence of childhood AT deficiency-associated severe liver disease indicates a genetic factor predisposing a subgroup of PiZ individuals to liver disease while others are protected in this respect[34,35] A lag in the degradation of the Z protein in the endoplasmic reticulum was indicated in genetically engineered fibroblast cell lines from PiZ individuals with liver disease compared with homozygotes without liver disease[36]. The exact biochemical and/or environmental aetiologies of the early severe progressive liver disease are not yet properly known.

In early childhood abnormal liver tests were recognized during routine check-ups in most of the children. When the blood samples were obtained in infected children the S-AT levels were often very high, some weeks later approaching or within the normal range again. These clinical observations have experimental correspondence. Liver injury involves a gain-of-toxic-function mechanism in which the endoplasmic reticulum retention of mutant Z protein triggers hepatotoxic reactions[37]. The Z protein folds slowly and may aggregate and form polymers[38]. Since AT is an acute-phase protein, more mutant protein is formed especially during bacterial infections, further exposing the endoplasmic reticulum to stress with a subsequent risk of liver cell death. The Z molecules are recognized by the protein control system directing them for proteolysis[20]. Mitochondria may be damaged, autophagy and apoptosis activated[21].

ADOLESCENCE AND YOUNG ADULTHOOD

A considerable decrease in the incidence of abnormal transferase concentrations was established during adolescence[38]. At age 18 years liver test abnormalities were found in 17% of PiZ subjects; none had clinical symptoms of liver disease. Only two individuals had abnormal tests both at 16 and 18 years of age. At age 26, 6% of PiZ subjects had marginally increased S-ALAT and 7% S-GT[5]. Again none had clinical symptoms of liver disease.

Lung function and clinical symptoms of lung disease have been followed since childhood. No obvious differences were found between PiZ, SZ and controls at 16–26 years of age[6,39,40]. Marginal deviations in complete spirometric studies were established at 18 years but no difference in FEV_1 or $FEV_1/FVC\%$ throughout young adulthood up to age 30 years (E. Bernspång, personal communication). However, the subgroups of 26-year-olds having suffered neonatal cholestasis or recurrent wheezers at age 2 years had significantly worse FEV_1 or $FEV_1/FVC\%$ at the 26-year follow-up[5]. When aged 26, 5% of the PiZ and SZ subjects were smokers.

Experimental studies by Rudnick and colleagues give us an insight into, and understanding of, the pathophysiology of liver abnormalities in the PiZ deficiency state[41]. Hepatocellular proliferation was studied in the liver of the PiZ transgenic mouse model. The liver cell proliferation was increased at baseline. This increase was proportional to the number of globule-containing liver cells and the level of Z protein accumulation within these cells[42]. However, the proliferating cells had no AT globules. Remarkably, only male mice had an increase of hepatocellular proliferation. In humans, males have outnumbered females by about 2:1 in neonates with cholestasis as well as in adults with cirrhosis and hepatocellular carcinoma[14,15]. The increase of cell death is quite low, as well as the number of proliferating cells, 2–3% detected over 72 h. The globule-containing liver cells are ‘sick but not dead’, with activated caspases, NF-κB and autophagy[21,27,42,43].

The pathophyisology of emphysema in AT deficiency is different from 'the gain of toxic function' in liver disease. By a 'loss of function' mechanism, the low AT level brings about a high risk of proteolytic damage to the connective tissue of the lung[4,44]. Cigarette smoking predisposes to severe lung disease in middle age[13,45]. The non-smoking population of AT-deficient now 30-year-old individuals have normal lung function despite the fact that the pathophysiological mechanisms said to cause emphysema in AT deficiency have been turned on since birth. During the first period of normal lung growth, elastogenesis is important to alveolar septation. The alveolar septa are composed of a capillary double layer with elastin localized at the tip[46]. Despite low AT, an inefficient immune system in early life and multiple viral and bacterial respiratory infections the lung growth seems to proceed normally. These contradictions may be explained by compensatory mechanisms. In our studies of protease inhibitors, elastase–AT complexes and neutrophil lipocalin (as a marker of neutrophil activity), compensatory changes were indeed indicated. In PiZ individuals the capacity of AT to bind elastase is reduced by 80–90%. The concentration of alpha-2-macroglobulin (A2M) is considerably higher during childhood and further increased in AT-deficient children and adolescents[6,47]. In AT deficiency considerably more elastase is in complex with A2M and these complexes down-regulate neutrophil proteases (*in-vitro* studies) as indicated in our observation of low neutrophil lipocalin levels in PiZ individuals[5,6]. At 26 years of age A2M was normal again, but at that age secretory leucocyte protease inhibitor (SLPI) was increased in AT-deficient subjects. In conclusion we hypothesize that increased concentrations of elastase–A2M complexes down-regulate the neutrophil secretion of proteolytic enzymes during childhood and adolescence. In young adulthood SLPI if upregulated, as indicated, may colocalize with neutrophils in the lung tissue and protect elastin fibres close to migrating neurtrophils.

PRESENT AND FUTURE THERAPEUTIC POSSIBILITIES

Today no effective treatment is available for AT deficiency and thus prevention has the highest priority. These high-risk individuals must, irrespective of whether lung or liver disease is considered, avoid a second 'hit' [48]. Smoking, environmental pollutants, suboptimally treated asthma, fasting, fatty liver disease, prolonged stimulation of acute-phase proteins and certain lung and liver infections are examples of additional hits or risks[49–51]. However, the common advice to avoid alcohol may be wrong[52].

In the group of diseases, nowadays designated conformational, AT deficiency is a prototype[2,53]. The Z mutant protein polymerizes before it is secreted, causing liver disease. This polymerization of the mutant protein can be prevented *in vitro* by specific peptides[54,55]. The function of the protein control system and the possibility of its manipulation are also intensively studied[56]. Gene therapy is also knocking at the door[57–59]. There is hope for a positive future for individuals suffering from conformational diseases.

References

1. Lomas DA, Mahadeva R. α_1-Antitrypsin polymerization and the serpinopathies: pathobiology and prospects for therapy. J Clin Invest. 2002;110:1585–90.
2. Carell RW, Lomas DA. Alpha-1-antitrypsin deficiency, a model for conformational diseases. N Engl J Med. 2002;346:45–53.
3. Perlmutter DH. Alpha-1-antitrypsin deficiency: diagnosis and treatment. Clin Liver Dis. 2004;8:39–59.
4. American Thoracic Society/European Respiratory Society Statement. Standards for the diagnosis and management of individuals with alpha-1-antitrypsin deficiency. Am Respir Crit Care Med. 2003;168:819–900.
5. Piitulainen E, Carlson J, Ohlsson K, Sveger T. α_1-Antitrypsin deficiency in 26-year-old subjects. Lung, liver and protease/protease inhibitor studies. Chest. 2005;128:2076–81.
6. Sveger T, Piitulainen E, Ohlsson K. Adolescents with α_1-antitrypsin deficiency have high α_2-macroglobulin and low neutrophil lipocalin and elastase levels in plasma. Pediatr Res. 1998;44:939–41,
7. Seersholm N, Kok-Jensen A. Clinical features and prognosis of life-time non-smokers with severe α_1-antitrypsin deficiency. Thorax. 1998;53:265–68.
8. Blanco I, de Serre FJ, Fernandez-Bustillo E et al. Estimated numbers and prevalence of PiS and PiZ alleles of α_1-antitrypsin deficiency in European countries. Eur Respir J. 2006;27: 77–84.
9. Beckman L. Sikström C, Mikelsaar A et al. α_1-antitrypsin (Pi) alleles as markers of West European influence in the Baltic Sea region. Hum Hered. 1999;49:52–5.
10. Thelin T, McNeil TF, Sveger T. Identifying children at high somatic risk (alpha-1-antitypsin deficiency): possible long term effects on parents' reproduction, marital status and social class level. Acta Psychiatr Scand. 1985;6:636–43.
11. Cox DW, Woo SL, Mansfield T. DNA restriction fragments associated with alpha-1-antitrypsin indicate a single origin for deficiency allele PiZ. Nature. 1985;316:79–81.
12. Lomas DA. The selective advantage of α_1-antitrypsin deficiency. Am J Respir Crit Care Med. 2006;173:1072–7.
13. Larsson C. Natural history and life expectancy in severe alpha-1-antitrypsin deficiency, PiZ. Acta Med Scand. 1978;204:345–51.
14. Eriksson S, Carlsson J, Velez R. Risk of cirrhosis and primary liver cancer in alpha-1-antitrypsin deficiency. N Engl J Med. 1986;314:736–9.
15. Sveger T. Liver disease in alpha-1-antitrypsin deficiency detected by screening of 200,000 infants. N Engl J Med. 1976;294:1316–21.
16. Oelberg DG. Prenatal growth: the sum of maternal, placental and foetal contributions. Pediatr Rev. 2006;27:224–8.
17. Malone M, Mieli-Vergani G, Mowat AP, Portman B. The foetal liver in PiZZ alpha-1-antitrypsin deficiency: a report of five cases. Pediatr Pathol. 1989;9:623–31.
18. Sveger T, Ekelund H. Variations of protease inhibitors in foetuses, newborn infants and in some neonatal disorders. Acta Pediatr Scand. 1975;64:763–9.
19. Lomas DA, Parfrey H. Alpha-1-antitrypsin deficiency. 4: Molecular pathophysiology. Thorax. 2004;59:529–35.
20. Sifers RN. Cell biology. Protein degradation unlocked. Science. 2003;299:1330–1.
21. Teckman JH, An JK, Blomenkamp K et al. Mitochondrial autophagy and injury in the liver in alpha-1-antitrypsin deficiency. Am J Physiol. 2004;286:G851–62.
22. Teckman JH, Lindblad D. Alpha-1-antitrypsin deficiency: diagnosis, pathophysiology and management. Curr Gastroent Rep. 2006:8:14–20.
23. Sveger T, Thelin T. Four-year-old children with alpha-1-antitrypsin deficiency. Acta Paediatr Scand. 1981;70:171–7.
24. Lester R. Physiol cholestasis. Gastroenterology. 1980:78:864–5.
25. Strandvik B, Wikström SÅ. Tetrahydroxylated bile acids in healthy human newborns. Eur J Clin Invest. 1982;12:301–5.
26. Janciauskiene S, Eriksson S. The interaction of hydrophobic bile acids with the α_1-proteinase inhibitor. FEBS Lett. 1994;343:141–5.
27. Teckman JH, An JK, Loethen S, Perlmutter DH. Fasting in α_1-antitrypsin deficient liver; constitutive activation of autophagy. Am J Gastrointest Liver Physiol. 2002;283:G1156–65.

28. Udall JN, Dixon M, Newman AP et al. Liver disease in α_1-antitrypsin deficiency. J Am Med Assoc. 1985;253:2679–82.
29. Sveger T. Breast-feeding, α_1-antitrypsin deficiency, and liver disease. J Am Med Assoc. 1985;254:30–6.
30. Field JC. The immunological components of human milk and their effect on immune development in infants. J Nutr. 2005;135:1–4.
31. Sveger T. Prospective study of children with α_1-antitrypsin deficiency: eight-year-old follow-up. J Pediatr. 1984;104:91–4.
32. Sveger T. Alpha-1-antitrypsin deficiency in early childhood. Pediatrics. 1978;62:22–25.
33. Perlmutter DH. Pathogenesis of chronic liver injury and hepatocellular carcinoma in alpha-1-antitrypsin deficiency. Pediatr Res. 2006;60:233–8.
34. Cox DW. Prenatal diagnosis of alpha-1-antitrypsin deficiency. Prenat Diagn. 2004;24:468–70.
35. Hinds R, Hadchouel A, Shanmugham NP et al. Variable degree of liver involvement in siblings with PiZZ alpha-1-antitrypsin deficiency-related liver disease. J Pediatr Gastroenterol Nutr. 2006;43:136–8.
36. Wu Y, Whitman I, Malmenti E et al. A lag in intracellular degradation of mutant α_1-antitrypsin correlates with the liver disease phenotype in homozygous α_1-antitrypsin deficiency. Proc Natl Acad Sci USA. 1994;91:9014–18.
37. Carlsson JA, Rogers BB, Sifers RN, et al. Accumulation of PiZ antitrypsin causes liver damage in transgenic mice. J Clin Invest. 1989;83:1183–90.
38. Sveger T, Eriksson S. The liver in adolescents with α_1-antitrypsin deficiency. Hepatology. 1995;22:514–17.
39. Sveger T, Piitulainen E, Arborelius Jr M. Lung function in adolescents with α_1-antitrypsin deficiency. Acta Pediatr. 1994;83:1170–3.
40. Sveger T, Piitulainen E, Arborelius Jr M. Clinical features and lung function in 18-year-old adolescents with α_1-antitrypsin deficiency. Acta Pediatr. 1995;84:815–16.
41. Rudnick DA, Perlmutter DH. Alpha-1-antitrypsin deficiency: a new paradigm for hepatocellular carcinoma in genetic liver disease. Hepatology. 2005;42:514–21.
42. Rudnick DA, Liao Y, An JK et al. Analysis of hepatocellular proliferation in a mouse model of α_1-antitrypsin deficiency. Hepatology. 2004;39:1048–55.
43. Teckman JH, Perlmutter DH. Retention of mutant α_1-antitrypsin Z in endoplasmic reticulum is associated with an autophagic response. Am J Physiol. 2000;279:G961–74.
44. Janoff A. Elastases and emphysema: current assessment of the antiprotease hypothesis. Am Rev Resp Dis. 1985;132:417–33.
45. Janus ED, Phillips NT, Carrell RW. Smoking, lung function and α_1-antitrypsin deficiency. Lancet. 1985;11:152–4.
46. Burri PH. Structural aspects of prenatal and postnatal development and growth of the lung. In: McDonald JA, editor. Lung Growth and Development. Marcel Dekker, New York, 1997:1–35.
47. Sveger T. Plasma protease inhibitors in α_1-antitrypsin deficient children. Pediatr Res. 1985; 19:834–5.
48. Parord ID, Birring SS, Berry M et al. Multiple inflammatory hits and the pathogenesis of severe airway disease. Eur Resp J. 2006;27:884–8.
49. Alfire ME, Treem WR. Non-alcoholic liver disease. Pediatr Ann. 2006;35:290–4, 297–9.
50. Brown A, Farmer K, Mac Donald L et al. House dust mite der p 1 down regulates defences of the lung by inactivating elastase inhibitors. Am J Respir Cell Mol Biol. 2003;29:381–9.
51. Sigsgaard T, Bonefeld-Jorgensen HJ, Hoffmann HJ et al. Microbial cell wall agents as occupational hazard. Toxicol Appl Pharmacol. 2005;207:s310–19.
52. Sharp LK, Mallya M, Kinghoarn KJ et al. Sugar and alcohol molecules provide a therapeutic strategy for the serpinopathies that cause dementia and cirrhosis. FEBS J. 2006;273:2540–52.
53. Carrell RW. Cell toxicity and conformational disease. Trends Cell Biol. 2005;15:574–80.
54. ParfreyH, Mahadewa NA, Ravenhill A et al. Targeting a surface cavity of alpha-1-antitrypsin to prevent conformational disease. J Biol Chem. 2003;278:33060–6.
55. Chang YP, Mahadeva R, Chang WS et al. Identification of a 4-mer peptide inhibitor that effectively blocks the polymerization of pathogenic Z α_1-antitrypsin. Am J Respir Cell Mol Biol. 2006;35:540–8.

56. Ying WU, Swulius MT, Moremer KW, Sifers RN. Elucidation of the molecular logic by which misfolded α_1-antitrypsin is preferentially selected for degradation. Proc Natl Acad Sci USA. 2003;100:8228–34.
57. Duan YY, Wu J, Zhu JL et al. Gene therapy for human alpha-1-antitrypsin deficiency in an animal model using SV40-derived vectors. Gastroenterology. 2004;127:1222–32.
58. Conlon TJ, Cossette T, Erger K et al. Efficient hepatic delivery and expression from a recombinant adenoassociated virus 8 pseudotyped alpha-1-antitrypsin vector. Mol Ther. 2005;12:867–75.
59. Park F, Gow KW. Gene therapy: future or flop. Pediatr Clin N Am. 2006;53:621–38.

11
Porphyrias

U. A. MEYER and A.-K. PEYER

INTRODUCTION

Porphyrias are rare metabolic disorders resulting from partial deficiency of enzymes of the haem biosynthesis pathway (Figure 1, Table 1). Depending on the location of the enzymatic defect within the pathway, different patterns of accumulation and excretion of intermediates, and accordingly two types of clinical symptoms, occur separately or in combination. Photosensitivity due to accumulation of porphyrins in skin is the main clinical manifestation of some porphyrias, whereas life-threatening attacks of neurovisceral symptoms occur in the acute inducible porphyrias named δ-aminolaevulinic acid dehydratase-deficient porphyria (ADP), acute intermittent porphyria (AIP), hereditary coproporphyria (HCP) and variegate porphyria (VP, Table 2). The attacks are precipitated by a number of well-known factors that increase the demand for haem such as drugs (e.g. barbiturates), alcohol, sex steroids or fasting. These attacks are always associated with increased activity of δ-aminolaevulinic acid synthase (ALAS1) in the liver and reflected by increased levels of the porphyrin precursors δ-aminolaevulinic acid (ALA) and porphobilinogen (PBG) in plasma and urine. Acute attacks are treated by withdrawal of the causative agent, infusion of glucose and of haem. If dehydration and imbalance of electrolytes ensue, attacks can be life-threatening and may require intensive care[1–3].

It is still under debate whether increased ALA concentration in the nervous system or a neuronal haem deficiency accounts for symptoms in acute attacks; most likely both mechanisms occur[4]. Liver transplantation has restored disrupted porphyrin metabolism in a symptomatic patient[5]. However, progress has been made in understanding the molecular mechanisms of ALAS1 regulation in the liver.

REGULATION OF ALAS1

Increased ALAS1 activity in the liver is the biochemical hallmark of acute attacks of the inducible hepatic porphyrias. ALAS1 activity is adjusted to the cellular needs by the end-product of the pathway, haem. Haem down-regulates

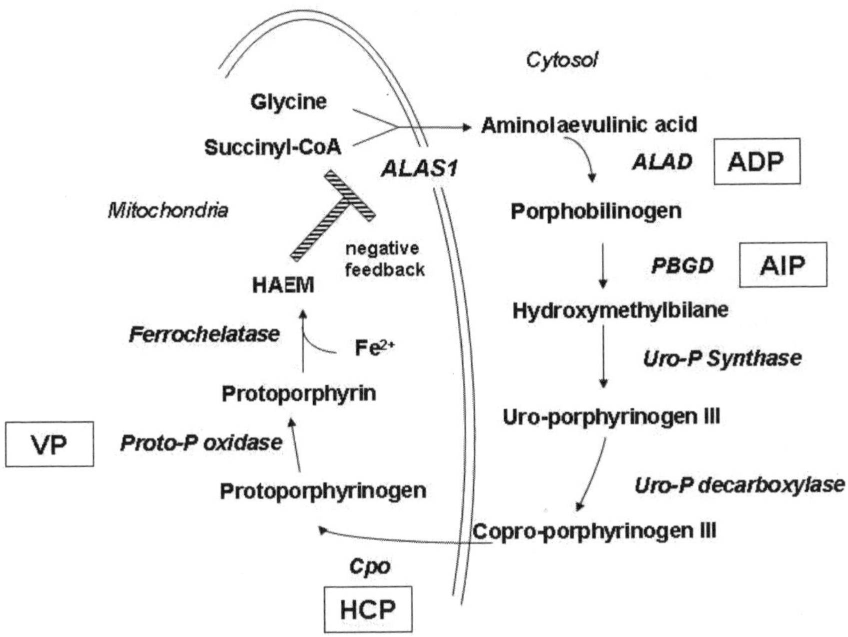

Figure 1 Haem synthesis pathway and inducible acute porphyrias. ALAS1, δ-aminolaevulinic acid synthase; ALAD, ALA dehydratase; PBGD, porphobilinogen deaminase; uro-P, uroporphyrinogen III; Cpo, coproporphyrinogen oxidase; Proto-P, protoporphyrinogen. ADP, ALA-dehydratase-deficient porphyria; AIP, acute intermittent porphyria; HCP, hereditary coproporphyria; VP, variegate porphyria. The bar between HAEM and ALAS1 indicates the negative feedback regulation of ALAS1 by haem

Table 1 Characteristics of the 4 inducible acute porphyrias (combined prevalence ~5 cases/100 000 persons)

Disease	*Inheritance*	*Deficient enzymes*	*Enzyme activity (%)*	*Known mutations*
AIP	Autosomal dominant	Porphobilinogen-l deaminase (HMBS*)	~50	>200
HCP	Autosomal dominant	Coproporphyrinogen oxidase	~50	>35
VP	Autosomal dominants	Protoporphyrinogen oxidase.	~50	>120
ADP	Autosomal recessive	ALA dehydratase (porphobilinogen synthase)	~5	7

HMBS, hydroxymethylbilane synthase.

Table 2 Symptoms and signs of inducible acute porphyrias (modified from ref. 1)

	Estimated incidence (%)
Gastrointestinal	
Abdominal pain	85–95
Vomiting	43–83
Constipation	48–84
Diarrhoea	5–12
Neuropsychiatric	
Pain in extremeties, back, chest, etc.	50–70
Paresis	42–68
Respiratory paralysis	9–20
Mental symptoms (agitation, confusion, hallucinations)	40–58
Convulsions	10–20
CV, autonomic nervous system	
Tachycardia	64–85
Arterial hypertension	36–55

transcriptions of ALAS1, shortens the half-life of ALAS1 mRNA and inhibits its import into the mitochondrial matrix[6]. Infusion of haem is the most effective way to treat acute attacks of porphyria[1,2].

Increased hepatic ALAS1 activity during acute attacks is the result of transcriptional activation of the ALAS1 gene, as all the precipitating factors such as drugs, hormones and fasting have been shown to affect ALAS1 transcription.

INDUCTION OF ALAS1 BY DRUGS

Studies in this laboratory have established that the transcriptional activation of the ALAS1 gene occurs by a mechanism similar to or identical to the induction of cytochromes P450, other drug-metabolizing enzymes and drug transporters. The 'xenosensing' nuclear hormone receptors CAR (constitutive androstane receptor) and PXR (pregnane X receptor) are the transcription factors activated by the inducer drugs. CAR and PXR dimerize with RXR (retinoid X receptor) and bind as transcription factor complex to cognate DNA response elements in the 5-flanking region of ALAS1[7–9]. In the presence of a partial deficiency of haem synthesis, as is the case in the inducible acute porphyrias, this induction response of ALAS1 to drugs is strikingly potentiated. This massive induction of ALAS1 was demonstrated in transgenic mice with a partial deficiency in haem synthesis developed in our laboratory[10,11].

INDUCTION OF ALAS1 BY FASTING

Whereas the induction of ALAS1 by drugs via CAR and PXR is well established, the response of the haem synthesis pathway to fasting and the beneficial effect of administration of glucose to symptomatic patients has remained enigmatic. In a recent collaborative study we could show that ALAS1 is regulated by the peroxisome proliferator-activated receptor γ-coactivator α (PGC-1α, Figure 2) via an insulin-sensitive FOXO1 site within the promoter of the ALAS1 gene[12]. PGC-1α is a coactivator of numerous nuclear receptors and transcription factors, and regulates mitochondrial biogenesis and gluconeogenesis. Its transcription is controlled by glucose availability. *In-vivo* studies in mice confirmed that ALAS1 is induced at the transcriptional level by fasting. Elevation of PGC-1α in mice via adenoviral vectors increased the levels of ALA and PBG in urine as observed in acute attacks of porphyria. The fasting effect was lost in liver-specific PGC-1α knockout animals, convincingly demonstrating that PGC-1α is the master regulator of the fasting response. Interestingly, induction of ALAS1 by drugs was not dependent on the presence of PGC-1α[12].

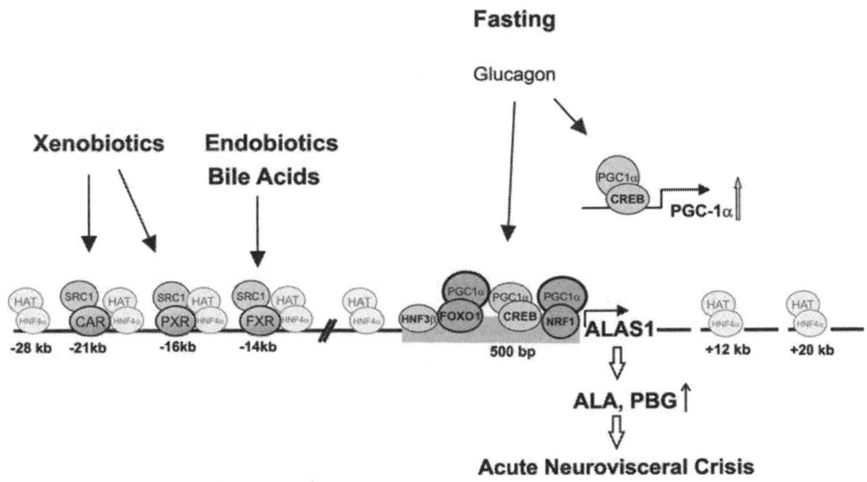

Figure 2 Working model of response of ALAS1 to xenobiotics and endobiotics via multiple nuclear receptors/transcription factors and coregulators. CAR and PXR activated by xenobiotics induce ALAS1 transcription via the two distal enhancer elements located at minus 21 kb and minus 16 kb respectively, in concert with coregulators such as steroid receptor coactivator 1 (SRC1) and PGC-1α (not shown). HNF4α binding sites interact with histone acetyltransferases (HAT) for liver-specific modulation of ALAS1 gene expression. Bile acid-activated FXR acts via the newly identified enhancer located at minus 14 kb. In the fasted liver a glucagon-triggered increase in cAMP directly stimulates ALAS1 transcription via the cAMP-responsive element binding element CREB, which also binds to the PGC-1α promoter[13]. The forkhead transcription factors FOXO1 and HNF3β (not shown) also bind to the ALAS1 promoter[12] and FOXO1 activity is enhanced by PGC-1α

ADDITIONAL FACTORS REGULATING ALAS1

The changing metabolic requirement for haem of liver cells suggests additional complex regulations of ALAS1. In an attempt to identify additional factors regulating ALAS1 we analysed by phylogenetic footprinting the genomic sequence of human ALAS1. At minus 14 kb of the transcriptional start site a novel response element for the bile acid receptor FXR, a nuclear receptor involved in lipid and glucose homeostasis as well as detoxification, was identified. Based on the molecular characterization of the FXR response element, and functional data in primary human hepatocytes and in mice, as well as on genomic sequence comparison, we could demonstrate that the response of ALAS1 to bile acid is unique to primates (Peyer et al., unpublished results, Figure 2).

In addition, multiple enhancer modules identified by cross-species sequence comparison were shown to be activated by hepatocyte nuclear factor 4α (HNF4α). HNF4α is potently activated by PGC-1α. Chromatin immunoprecipitation confirmed the binding of this factor in a native chromatin context (Peyer et al., unpublished results, Figure 2). We therefore hypothesize that the liver-enriched transcription factor HNF4α is required for basal expression and liver-specific induction of ALAS1.

The clinical relevance of bile acid-mediated induction of ALAS1, i.e. whether increased bile acids can provoke acute attacks of inducible hepatic porphyrias via FXR, is not known. Our inquiries to centres of porphyria research in Scandinavia and South Africa have not answered this question, but have made these physicians aware of this possibility. Similarly, the physiological role of HNF4α as a regulator of ALAS1 is not yet clear.

CONCLUSIONS

The mechanisms which trigger acute neurovisceral symptoms of inducible hepatic porphyrias have been evaluated. Induction of ALAS1 by drugs which induce cytochromes P450 is now well understood as mediated by the xenosensors CAR and PXR. We now also understand the effect of fasting on haem synthesis and the beneficial effect of glucose as mediated via PGC-1α. The importance of additional regulatory effects of FXR and HNF4α in the complex regulation of the transcription of the ALAS1 gene, however, remain less clear.

Acknowledgements

This research was supported by the Swiss National Research Foundation. A.-K. Peyer was supported by a MD/PhD fellowship of the Roche Research Foundation.

References

1. Anderson KE, Bloomer JR, Bonkovsky HL et al. Recommendations for the diagnosis and treatment of the acute porphyrias. Ann Intern Med. 2005;142:439–50.
2. Kauppinen R. Porphyrias. Lancet 2005;365:241–52.
3. Badminton MN, Elder GH. Molecular mechanisms of dominant expression in porphyria. J Inherit Metab Dis. 2005;28:277–86.
4. Meyer UA, Schuurmans MM, Lindberg RLP. Acute porphyrias: pathogenesis of neurological manifestations. Semin Liver Dis. 1998;18:43–52.
5. Soonawalla ZF, Orug T, Badminton MN et al. Liver transplantation as a cure for acute intermittent porphyria. Lancet. 2004;363:705–6.
6. May BK, Dogra SC, Sadlon TJ, Bhasker CR, Cox TC, Bottomley SS. Molecular regulation of heme biosynthesis in higher vertebrates. Prog Nucl Acid Res Mol Biol. 1995;51:1–51.
7. Fraser DJ, Podvinec M, Kaufmann MR, Meyer UA. Drugs mediate the transcriptional activation of the 5-aminolevulinic acid synthase (ALAS1) gene via the chicken xenobiotic-sensing nuclear receptor (CXR). J Biol Chem. 2002;277:34717–26.
8. Fraser DJ, Zumsteg A, Meyer UA. Nuclear receptors constitutive androstane receptor and pregnane X receptor activate a drug-responsive enhancer of the murine 5-aminolevulinic acid synthase gene. J Biol Chem. 2003;278:39392–401.
9. Podvinec M, Handschin C, Looser R, Meyer UA. Identification of the xenosensors regulating human 5-aminolevulinate synthase. Proc Natl Acad Sci USA. 2004;101:9127–32.
10. Lindberg RLP, Porcher C, Grandchamp B et al. Porphobilinogen deaminase deficiency in mice causes a neuropathy resembling that of human hepatic porphyria. Nat Genet. 1996;12: 195–9.
11. Lindberg RLP, Martini R, Baumgartner M et al. Motor neoropathy in porphobilinogen deaminase-deficient mice imitates the peripheral neuropathy of human acute porphyria. J Clin Invest. 1999;103:1127–34.
12. Handschin C, Lin J, Rhee J et al. Nutritional regulation of hepatic heme biosynthesis and porphyria through PGC-1alpha. Cell. 2005;122:505–15.
13. Yoon JC, Puigserver P, Chen G et al. Control of hepatic gluconeogenesis through the transcriptional coactivator PGC-1. Nature. 2001;413:131–8.

12
Liver disease in cystic fibrosis

C. COLOMBO, S. BOTTO POALA, V. MOTTA and L. ZAZZERON

INTRODUCTION

Cystic fibrosis (CF) is the most common life-limiting autosomal recessive disease of the Caucasian population; with an incidence of approximately 1 in every 3000 live births worldwide[1]. The disease affects several organs, more frequently the lungs, pancreas, sweat glands and in males the Wolffian ducts. Lung disease is the primary cause of morbidity and mortality and results from the combined and sustained effect of chronic infection from different respiratory pathogens and inflammation leading to progressive damage and eventually to respiratory failure.

When first described in 1938 the disease was almost invariably fatal during early childhood, and for many years the basic defect has remained unknown. To date, median survival approaches the age of 40 years, but premature death due to respiratory failure is still a major problem[2].

In 1989 the discovery of the gene responsible for CF[3], led to recognition of the key role of the encoded protein, the cystic fibrosis transmembrane regulator (CFTR), in maintaining fluid balance across epithelial cells. CFTR, a large protein of 1480 amino acids, belongs to the ATP-binding cassette family and functions mainly as a cAMP-dependent chloride channel at the apical membrane of the epithelial cell where it promotes transmembrane efflux of chloride ions[4]. The CF secretory defect determines inability to maintain the luminal hydration of ducts, leading to physicochemical abnormalities of secretions and duct obstruction.

CF is characterized by a striking genetic heterogeneity, with more than 1400 different disease-causing mutations identified to date[5]. Mutations are distinguished as severe and mild on the basis of the presumed molecular mechanisms of dysfunction of the CFTR protein and its residual activity at the apical membrane of epithelial cells[6].

The phenotypic expression of CF disease is also extremely heterogeneous in terms of severity and type of organs involved: while the sweat glands, pancreas, lungs and Wolffian ducts in male subjects are affected in the majority of patients, the liver and the intestine are less frequently involved.

The major clinical manifestations of CF, the pathogenetic mechanism and relative frequencies are reported in Table 1.

Table 1 Major clinical manifestations in cystic fibrosis

Organ involved	*Functional/structural abnormalities*	*Clinical manifestation*	*Frequency*
Sweat glands	Salt hyperconcentration	Elevated sweat chloride concentrations	95–98%
		Heat prostration shock	Undefined
Intestine	Thickened intestine content	Meconium ileus	10–20%
		Distal intestinal obstruction	10%
		Rectal prolapse	1–2%
Liver and biliary tract	Thickened ductular secretions	Neonatal cholestasis	Rare
	Lithogenic bile	Microgallbladder	30%
		Cholelithiasis	15%
		Focal biliary cirrhosis	20–30%
		Multilobular biliary cirrhosis	10%
		Portal hypertension	2–5%
		Liver steatosis	23–67%
		Liver enzyme alterations	40–70%
		Sclerosing cholangitis	Rare
		Common bile duct stenosis	Rare
Pancreas	Pancreatic fibrosis	Pancreatic insufficiency	85%
	Ducts obstruction with fibrocystic transformation		
	Inflammatory response in the surviving pancreatic tissue	Chronic recurrent pancreatitis (if pancreatic sufficiency)	3–4%
	Islet cell strangulation	Diabetes mellitus	10–25%
Nose and paranasal	Mucus stagnation/infection	Sinusitis	90%
		Mucocoele	16%
		Nasal polyps	15–26%
Lung	Defective clearance of mucus–obstruction–infection–inflammation	Chronic respiratory infections	97%
		Pneumothorax	5–8%
		Massive haemoptysis	5%
		Allergic bronchopulmonary aspergillosis	10%
Reproductive system	Atresia–azoospermia	Male sterility	98%
	Mucus abnormalities	Female decreased fertility	Undefined

Although a relation between genotype and phenotype has been established only for pancreatic status[6,7], the term 'classic CF' refers to the more severe form of the disease, with multiorgan involvement and pancreatic insufficiency, and is associated with the presence of two severe mutations. Non-classic forms of CF occur in 10% of cases, are associated with mild mutations and are characterized by residual pancreatic function and often monosymptomatic disease (chronic bronchitis, sinusitis with nasal polyposis, pancreatitis, male infertility due to obstructive azoospermia)[8].

SPECTRUM OF HEPATIC MANIFESTATION

Over the past two decades, with improved survival of CF patients, increased attention has been devoted to extrapulmonary manifestations of the disease, particularly CF-associated liver disease, CF-related diabetes and infertility.

Within the hepatobiliary system a wide spectrum of clinical problems has been recognized that include specific alterations ascribable to the underlying CFTR defect, as well as lesions of iatrogenic origin or reflecting the effects of a disease process that occurs outside the liver[9] (Table 2).

Table 2 Major hepatic manifestations in CF

Type of lesion	*Clinical manifestation*	*Frequency*
Specific alterations ascribable to the underlying CFTR defect	Focal biliary cirrhosis	20–30%
	Multilobular biliary cirrhosis	10%
	Portal hypertention	2–5%
	Neonatal cholestasis	Rare
	Sclerosing cholangitis	Rare
	Microgallbladder	30%
	Cholelithiasis	15%
Lesions of iatrogenic origin	Liver steatosis	23–67%
	Drug hepatotoxicity	Undefined
Lesions reflecting the effects of a disease process that occurs outside the liver	Hepatic congestion	Rare
	Common bile duct stenosis	Rare

Focal biliary cirrhosis leading to multilobular cirrhosis and portal hypertension is the classic and pathognomonic hepatic lesion of CF; it is considered a direct consequence of the basic defect and will be the main object of this review.

Although rarely, neonatal cholestatic jaundice may be the first manifestation of CF, and results from obstruction of extrahepatic bile ducts by viscous biliary secretions. This is also a direct consequence of the basic CF defect. Cholestasis may be complete, mimicking biliary atresia[10], but generally resolves spontaneously within the first months of life and does not progress to clinically relevant liver disease. At greater risk are infants with meconium ileus, particularly those with other conditions enhancing the risk of cholestasis, i.e. total parenteral nutrition or abdominal surgery.

Hepatic steatosis has been reported in CF patients of any age, with prevalence figures ranging between 23% and 67%. The pathogenesis is still largely unknown, but does not seem to be directly related to the CF basic defect. Massive steatosis, once frequently observed in newly diagnosed patients with pancreatic insufficiency and severe malnutrition, has become infrequent due to earlier diagnosis and to more appropriate nutritional care. Steatosis of milder severity has been associated with selective nutritional deficiencies, including essential fatty acids[11], carnitine, and more recently choline due to altered phospholipid metabolism in CF[12]; it may also be a consequence of diabetes or long-term antibiotic therapy. It has been suggested that, in CF patients chronically colonized by respiratory pathogens, steatosis may result from the effect of circulating cytokines on hepatic fatty acid oxidation or mitochondrial function[13]. Steatosis has so far been considered a benign condition, without a proven relationship to the subsequent development of cirrhosis; however, in a few patients steatosis was recognized as the first step in the progression towards more severe hepatic lesions. Studies in progress on the role of non-alcoholic steatohepatitis as a cause of cirrhosis in adults[14], may lead us to reconsider this issue in the future.

Stenosis of the intrapancreatic portion of the common bile duct due to pancreatic fibrosis is another example of hepatic damage due to the effect of a disease process occurring outside the liver. This is a rare complication that may cause complete or partial biliary obstruction, with jaundice and/or biliary pain, and may respond to surgical decompression[15].

Sclerosing cholangitis was described in a few adult CF patients, sometimes in association with inflammatory bowel disease[16]; however, some evidence of abnormalities of the biliary tree resembling sclerosing cholangitis can be obtained in the majority of CF patients with liver disease[17].

THE FOCAL BILIARY CIRRHOSIS/MULTILOBULAR CIRRHOSIS CONTINUUM

As previously mentioned, the typical hepatic manifestation of CF is focal biliary cirrhosis, which results from biliary obstruction and progressive periportal fibrosis. This is the most clinically relevant CF-associated hepatic problem, since extension of the initially focal fibrogenic process may lead to multilobular biliary cirrhosis, portal hypertension and related complications[18]. Unlike pulmonary and pancreatic diseases, that affect around 90% of CF patients, liver disease develops in no more than one-third of CF patients. Nevertheless, liver disease and liver failure remain the single most important non-pulmonary causes of death, accounting for 2.5% of overall CF mortality[19].

Pathogenesis

CF-associated liver disease is currently considered the first example of inherited liver disease resulting from impaired secretory function of the biliary epithelium; ductal cholestasis due to reduced fluid and electrolyte transport by cholangiocytes is considered the central step in the pathogenetic sequence of CF-associated liver disease.

In the hepatobiliary system, CFTR is expressed at the apical membrane of cholangiocytes and gallbladder epithelial cells, but not in hepatocytes[20]. At this level CFTR regulates the fluid and electrolyte content of bile. In physiological conditions the CAMP-stimulted Cl^- secretion through the low-conductance Cl^- channel imposes a negative luminal potential and an osmotic gradient that triggers passive secretion of Na^+ and water. The change in apical Cl^- gradient facilitates HCO_3^- extrusion via Cl^-/HCO_3^- exchange, providing the biliary alkalinization required for the digestive function and for the solubility of organic components of bile. In CF patients the absence or dysfunction of CFTR may reduce bile fluidity and alkalinity, leading to plugging of intrahepatic bile ducts by inspissated biliary secretions.

Injury of bile duct cells seems to represent the primary event in the development of periportal fibrosis, as suggested by the ultrastructural abnormalities of cholangiocytes consistently documented in CF patients, with irregular shapes, necrosis and periductular collagen deposition[21]. Quantitative and qualitative abnormality in mucin secretion may also contribute to abnormal bile viscosity in CF patients: secretion of chondroitin sulphate was shown to be markedly increased in CF biliary epithelium *in vitro*[22], and its accumulation may well explain bile duct plugging by eosinophilic material, which is one of the early histological changes found in infants and children with CF[23].

Altered bile composition and/or decreased bile flow cause obstruction of small bile ductules, which is followed by retention of hydrophobic bile acids, secondary hepatocyte injury with release of proinflammatory cytokines, growth factors, or lipid peroxide products that recruit and activate hepatic stellate cells to synthesize collagen[18].

In CF the focal inspissation of biliary secretions in intrahepatic bile ducts leads gradually to the development of portal fibrosis, bridging and, eventually, cirrhosis. This process begins focally in the liver, possibly because of interductal connections that may allow adequate bile drainage from some areas[24]. The progression from cholestasis to focal biliary cirrhosis to multilobular cirrhosis may take years to decades, and should be viewed as a continuum.

In summary, current evidence suggests that liver disease in CF is related to the CF basic defect at the hepatobiliary level. However, it remains to be explained why only one-third of CF patients develop liver disease and why liver disease shows a great degree of variability in terms of severity. Although development of liver disease seems to be restricted to patients with a severe genotype, no specific CFTR mutations has been associated with the presence and severity of liver disease[25,26], suggesting a multifactorial pathogenesis. The role of environmental factors is suggested by the occasional familial clustering of liver disease, and these may include factors related to poor nutritional status, antioxidant deficiency, poor compliance with treatment or chronic drug hepatotoxicity. On the other hand, a study of siblings with discordant liver expression has suggested that genetic factors inherited independently from the CFTR gene could modulate the clinical expression and severity of liver disease in CF[27]. The role of modifier genes in CF is the subject of much current research interest[28]: preliminary data obtained through a multicentre study suggest a complex multigenic inheritance for CF-associated liver disease, with possible interaction of different genes (α_1 anti-trypsin deficiency, transforming growth factor beta cytokine, mannose-binding lectin 2 and glutathione *S*-transferase)[29]. Identification of genetic modifiers is a priority since it may allow identification of patients at risk for the development of liver disease at the time of diagnosis of CF and early institution of prophylactic strategies.

Prevalence of liver disease

There are marked differences in the reported prevalence of CF-associated liver disease which may be explained by the use of different diagnostic criteria and on the population studied. Data derived from autopsy studies have clearly documented a progressive increase in prevalence of liver disease with age, from 10% in infants[23] to 72% in adults[30].

Prevalence seems to increase through childhood into mid-adolescence, with no significant increase thereafter. This suggests that, for CF patients who will develop liver disease, the mechanism and risk factors for liver damage are already present in early childhood. Higher prevalence figures have been reported in studies in which liver disease was actively searched for and diagnosed through a combination of diagnostic tools (clinical, biochemical, ultrasonographic evaluation)[31].

Table 3 Evolution of liver disease in CF

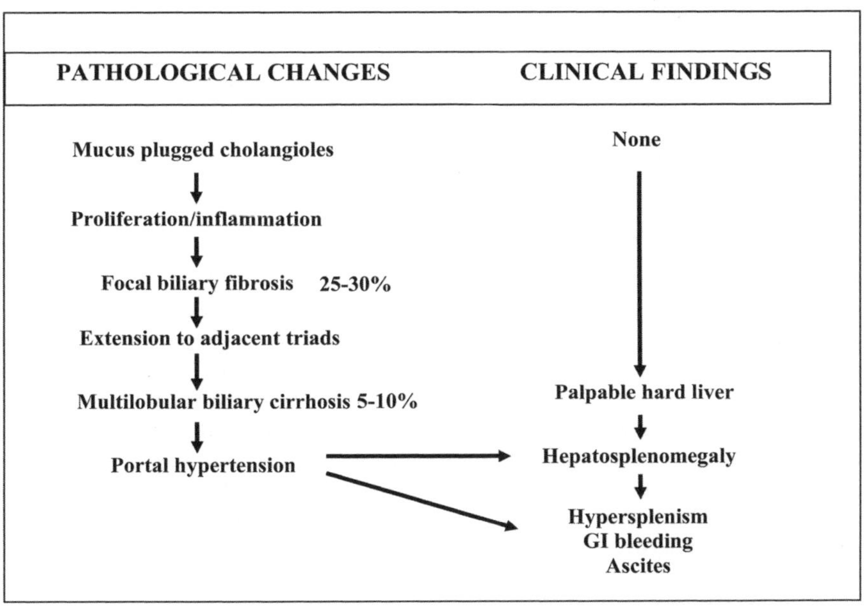

Natural history and clinical features

Patients are initially asymptomatic; jaundice is limited to patients with end-stage liver disease. Multilobular biliary cirrhosis is considered to develop sequentially from focal biliary cirrhosis; however, this progression occurs only in a minority of patients and the reasons for this progression are unknown. The relevance of degree of abnormalities in liver biochemistry or detection of non-specific ultrasonography findings (fatty infiltration, increased liver texture) in the development of severe liver disease needs to be explored by prospective studies.

As in other liver diseases characterized by initial involvement of bile ducts and not of hepatocytes, liver failure is a late event. In contrast, the haemodynamic consequences of cirrhosis are often prominent, favouring early development of portal hypertension[32,33].

Progression of liver disease from early stage to liver failure may take decades to occur, although in a minority of patients, often in the paediatric age group, liver disease may represent the main clinical problem, and its progression may be unusually rapid.

In a retrospective study spanning a 26-year period, of 44 children with CF investigated for liver cirrhosis, oesophageal varices developed in 86%, and in half of these patients oesophageal bleeding occurred early during the second decade of life. Liver failure occurred in 36% of patients at a mean age of 15 years[32]. More recently a median survival of 8.4 years has been reported in a group of CF patients after variceal bleeding as compared to a 1-year survival of only 34% in the general cirrhotic population, suggesting that a history of bleeding in the absence of decompensated cirrhosis may not represent an indicator of a markedly adverse prognosis in CF[34].

Incidence and outcome of liver disease in CF patients has been evaluated also through prospective studies in carefully monitored CF patients[11,35,36]. Long-term follow-up of different cohorts of CF patients carefully monitored for hepatic status has indicated cumulative incidence of liver disease ranging between 27% and 35%, without incident cases after the age of 18 years[11,36].

In these studies liver disease has shown a relatively slow progression, that, in a subgroup of patients reported by Lindblad et al., was confirmed by histologic data[11].

In our study[36], during more than 8 years from diagnosis of liver disease, the incidence of major complications of cirrhosis (oesophageal varices, gastrointestinal bleeding, ascites) was 0.4 cases per 100 cirrhotic patient-years (95% CI 0–2.0) (Table 4). Hepatic synthetic function was sufficiently maintained in the majority of cirrhotic patients, and only one liver transplantation was required[36]. The impact of liver disease on the outcome of CF was negligible: no significant differences between patients who developed and those who did not develop liver disease were observed in rate of respiratory failure, need for oxygen therapy or number of hospitalizations during the last 2 years of follow-up, suggesting that liver disease does not expose CF patients to higher risk of severe pulmonary disease or other major outcome events. Mortality from any cause also did not differ significantly among patients with different liver status, indicating lack of association between liver disease and mortality in CF.

Table 4 Incidence of liver disease and its complications in CF (and 95% CI)

	Incidence: No. per 100 patient-years
Liver disease	
Overall	1.8 (1.3–2.4)
First 10 years of life	2.5 (1.8–3.3)
Cirrhosis, liver disease patients	4.5 (2.3–7.8)
Portal hypertension, cirrhotic patients	28.8 (15.4–49.3)
Liver decompensation, cirrhotic patients	0.4 (0–2.0)
Death, any cause, or orthotopic liver transplant	
Liver disease patients	0.4 (0.1–1.2)
Cirrhotic patients	1.6 (0.3–4.7)

In summary, liver disease associated with CF develops in a significant percentage of CF patients, usually before or at puberty, and displays a slowly progressive course. Until the most advanced stages are attained, liver disease does not seem to influence nutritional status and severity of pulmonary involvement.

Impact of end-stage liver disease in CF

In contrast, the impact of advanced liver disease on the pulmonary function and the nutritional status of CF patients is becoming increasingly evident. During the progression of liver disease, CF patients are at risk of developing not only complications related to portal hypertension, but also several extrahepatic complications, including malnutrition, hepatic osteodystrophy and deterioration of pulmonary status.

The pathogenesis of malnutrition is multifactorial and involves the increase in resting energy expenditure (REE) characteristic of the disease, abnormalities in nutrient intake (as a result of anorexia and, in a few patients, to protein restriction), malabsorption (due to the combined effect of cholestasis and pancreatic insufficiency) and abnormal metabolism of nutrients[37] (Table 5). Many CF patients begin to show decompensated liver disease as adolescents, when relative glucose intolerance may develop due to latent diabetes mellitus; in addition advanced liver disease may induce insulin resistance, and thus represent a major risk factor for the development of CF-related diabetes[38].

CF patients with liver disease are also at risk of developing osteoporosis. In a study aimed at establishing the effects of end-stage liver disease on body composition we have documented severe osteopenia in a group of patients with multilobular cirrhosis and portal hypertension, that was significantly corrected after transplantation but not after conservative management[39]. The effect is presumably linked to restoration of hepatic function and bile flow after transplantation.

Table 5 Causes of malnutrition in CF patients with liver disease

Increased resting energy expenditure (REE)
Reduced caloric intake
Anorexia
Protein restriction
Malabsorption of lipid-soluble nutrients/vitamins
Reduced bile flow
Pancreatic insufficiency
Abnormal metabolism of nutrients:
Carbohydrates: glucose intolerance/diabetes, reduced hepatic glycogen
Protein: hypercatabolic state, increased tyrosine and phenylalanine, decreased branched amino acids
Lipids: essential fatty acid deficiency
Vitamin D: reduced 25-hydroxylation
Vitamin A: defective release from liver stores

With regard to pulmonary status, cirrhosis and portal hypertension can negatively affect respiratory function due to organomegaly, ascites-induced diaphragmatic splinting and intrapulmonary shunting, leading to recurrent respiratory infections from multiresistant bacteria, frequent hospital admissions and significant deterioration of quality of life.

Risk factors for liver disease in CF

Several factors have been recognized to be significantly associated with the development of CF-liver disease, including male sex, age at diagnosis, history of meconium ileus, pancreatic insufficiency and severe genotype[25,26,36,40,41], however, their role could not be consistently confirmed in different studies.

The preponderance of male subjects among CF patients with liver disease has been reported by several investigators[32,33,36,42] and more recently confirmed by a survey of European CF patients who had undergone liver transplantation[43]. Male predominance suggests that endocrine factors may exert a protective effect on female CF patients, who in our prospectively followed cohort did not develop liver disease after the age of 11 years[36].

We also found that the incidence of liver disease was significantly higher in patients with a positive history of meconium ileus, most of whom had developed liver disease earlier during the first decade of life. This association, linking inspissated gut content and biliary secretions, was first described by a necropsy study[44], but was not consistently found in cross-sectional studies involving different CF patient populations[11,26], and a study recently reported that meconium ileus was even associated with absence of liver disease[41]. In patients with a positive history of meconium ileus, additional risk factors for development of liver disease may include abdominal surgery with extensive small bowel resection, poor nutrition in early life and prolonged total parenteral nutrition[10]. Alternatively, a number of still-unknown factors, including genetic modifiers associated with gender or ethnic group, may confer CF patients different susceptibility to liver involvement.

With regard to genotype, despite the already-mentioned lack of association between any specific CFTR mutation and the development of liver disease, the incidence of liver disease has been generally found to be higher in patients carrying severe genotypes. The presence of a severe mutation seems therefore to represent a necessary but not a sufficient condition for the development of liver disease in CF.

More recently one study from Ireland reported that age at diagnosis may represent an important risk factor for development of liver disease[41]: the finding that later diagnosis resulting in poor nutrition status may predispose CF children to liver disease, supports further recommendations for implementation of newborn screening programmes aimed at earlier diagnosis of this condition[45].

Diagnosis of CF-associated liver disease

Evidence of liver disease in CF patients is often subclinical until pathological changes are advanced; therefore it is often underdiagnosed and early detection can be difficult. The most common presentation is the occasional finding of hepatomegaly on routine physical examination, often but not always associated with abnormalities of liver biochemistry.

Biochemical abnormalities are frequently mild or intermittently present and have shown low sensitivity and no correlation with histological findings[46]; not infrequently CF patients with multilobular biliary cirrhosis have completely normal liver biochemistry. In addition, non-specific biochemical abnormalities have been documented in more than 50% of infants with CF, with complete normalization in most cases within 2–3 years of age and no impact on future development of liver disease[11]. Isolated elevation of serum transaminase levels, with normal concentrations of enzyme related to cholestasis (gamma-glutamyl transpeptidase and alkaline phosphatase), suggests the presence of steatosis, which should be adequately recognized and followed up, after correction of nutritional deficiencies, if present.

Histological assessment, which represents the gold standard in the diagnostic work-up of many chronic liver diseases, may under-represent the extent of the disease due to patchy distribution of lesions[46]. Nevertheless, liver biopsy may provide important information on the type of predominant lesion (steatosis or focal biliary cirrhosis), the extent of portal fibrosis, rate of progression of liver disease and response to treatment. It should probably be performed more frequently, particularly if there is doubt about the diagnosis, and also to establish cirrhosis or prior to treatment with ursodeoxycholic acid or liver transplantation.

At present the diagnosis of CF-associated liver disease is still based on regular clinical examination (with measurement of liver span at the mid-clavicular line) and on a combination of biochemical tests and imaging techniques, particularly ultrasonography.

Ultrasound scan of the hepatobiliary system is recommended as the most suitable initial non-invasive method of investigation. Ultrasound technology has improved dramatically in recent years and can now reliably distinguish different patterns of the disease such as normal parenchyma, fatty infiltration, fibrosis, cirrhosis, portal hypertension or ductal abnormalities. Abnormal echogenicity frequently precedes clinical and biochemical manifestations of liver disease, suggesting that routine ultrasonography may be a valuable marker of early liver disease in CF[47,48]. To overcome the problem of inter- and intra-observer variability, attempts have been made to standardize the method of assessment, and scoring systems have been developed. A simple echographic scoring system based on coarseness of liver parenchyma, nodularity of the liver edge and increased periportal echogenicity has been proposed for the hepatic follow-up of these patients[49]. More recently, Stewart proposed an alternative scoring system that evaluates many different components of liver disease, but its validity is still to be evaluated[50].

Hepatobiliary scintigraphy with iminodiacetic acid derivatives can document a typical picture of biliary drainage impairment, with dilation of intrahepatic and extrahepatic bile ducts and delayed biliary excretion and intestinal appearance of the tracer[51]. Scintigraphy also provides functional information and has been employed to document time-related progression of liver disease[52] and to monitor the response to treatment with ursodeoxycholic acid[53].

Endoscopic retrograde cholangiography (ERCP) is an invasive procedure with considerable potential for complications. It remains the investigation of choice for patients with symptoms and signs suggestive of sclerosing cholangitis[16], distal stenosis of the common bile duct[15] and for those with choledocholithiasis, for which it also provides a therapeutic option.

More recently magnetic resonance cholangiography imaging of the biliary tree in CF patients has been obtained. Cholangitic lesions were detected in all patients with liver disease and in half of those without clinically apparent liver disease, thus suggesting that such a technique may be employed for early detection of intrahepatic biliary tract involvement, which is not achievable with other non-invasive methods[17].

Hepatic follow-up of CF patients

Regular monitoring of hepatic status should be accomplished in all patients with CF, with liver biochemistry and ultrasound scan included in their routine annual schedule. In patients with cirrhosis it seems prudent to determinate α-fetoprotein levels annually to monitor the possible development of hepatocellular carcinoma[54].

Management of CF-associated liver disease

Because of the decreasing mortality from extrahepatic causes, management of liver disease in CF patients is becoming a relevant clinical issue. With better knowledge of mechanisms involved in the pathogenesis there is potential for several therapeutic strategies, including liver-targeted gene therapy and pharmacological correction of the defective ion transport function.

Somatic gene therapy aimed at replacing the defective gene to the biliary epithelium has been shown to be feasible in the experimental animals[55], and may be curative. However, cholangiocytes are less easily accessible to targeted drug and gene delivery than the airways epithelium, and the clinical application of this approach remains to be established.

Strategies aimed at stimulating fluid secretion by cholangiocytes may be a useful approach to treat CF-associated liver disease, where reduced Cl^- and fluid secretion causes inspissated bile and chronic liver damage. Extracellular adenosine 5′-triphosphate (ATP) is known to be a potent Cl^- secretagogue and can activate the Ca^{2+}-dependent Cl^- channel in different cell types, including biliary epithelial cells[56]. Exogenous administration of nucleotide triphosphate may therefore correct the defective anion secretion in intrahepatic ducts and represent a potential pharmacological approach to CF-associated liver disease.

The ability of glibenclamide, a sulphonylurea with known CFTR inhibitor activity, to paradoxically stimulate cholangiocyte secretion through exocytosis in CFTR defective mice, has recently suggested that drugs with similar mechanism of action may be designed for the treatment of CF-associated and other cholestatic liver diseases[57].

Bile acid therapy

Oral bile acid therapy, aimed at improving biliary secretion in terms of bile viscosity and bile acid composition, is currently the only available therapeutic approach for CF-associated liver disease.

Ursodeoxycholic acid (UDCA) administration can contrast several of the mechanisms involved in cholestasis-induced liver injury[58]. The changes induced on bile acid composition through replacement and/or displacement of detergent, hydrophobic endogenous bile acids would prevent perpetuation of liver damage caused by their retention within the hepatocyte during cholestasis.

Other mechanisms of action may be involved, including direct cytoprotection of UDCA on biological membranes, a protective effect against apoptosis induced by endogenous bile acids and stimulation of bile secretion by hepatocytes and bile duct epithelial cells[58]. Recent data indicate that the beneficial effects of UDCA in CF-asociated liver disease may be mainly related to stimulation of Cl^- secretion through Ca^{2+}-dependent Cl^- conductance and to concomitant reduction in mucin secretion[59].

The optimal daily dose of UDCA has been established to be 20 mg/kg body weight[60]. This dose is higher than that conventionally used in other cholestatic liver diseases and this is probably due to poor intestinal bile acid absorption in CF patients.

Although no systematic study dealing with UDCA safety in CF-associated liver disease and other paediatric liver diseases has so far been reported in the literature, UDCA is generally well tolerated without significant adverse effects. The benefical effects on liver biochemistry have been consistently documented, also in the context of randomized clinical trials[61]. There is also strong evidence of improvement in liver histology[62] and in essential fatty acid status[63]. The latter may have potential extrahepatic implications since essential fatty acid deficiency seems to occur in CF from a specific defect in fatty acid metabolism producing a marked membrane lipid imbalance, which in turn may influence inflammatory status and the severity of pulmonary disease[64]. The beneficial effects of UDCA on hepatic excretory function, biliary drainage[53], quantitative liver function and nutritional status have been reported only by uncontrolled trials, and one study reported that UDCA led to normalization of early changes at ultrasonography[65].

No data are available on clinically relevant events or on survival[66]; however, in consideration of the long natural history and variable time-course of CF-associated liver disease, randomized clinical trials, with high numbers of patients followed up for many years, would be required to demonstrate the long-term efficacy of UDCA.

UDCA is currently widely used for the treatment of CF-associated liver disease and it is unlikely that long-term randomized studies with meaningful numbers of patients will ever be available. Asymptomatic patients with early-stage liver disease are more likely to benefit from UDCA administration. Identification of CF patients who are at risk for its development is also needed to evaluate if UDCA also has a role for the prevention of liver disease.

Treatment of portal hypertension and end-stage liver disease

In CF patients with more advanced liver disease, severe portal hypertension and hypersplenism are the most relevant problems. Variceal bleeding may require sclerotherapy or preferably variceal ligation during the acute episode: endoscopic treatment is successful in most cases and, in the absence of decompensated cirrhosis, can usually be managed without liver replacement[34].

As previously mentioned, CF patients with severe portal hypertension and hypersplenism may remain in stable conditions for years, and long-term survival has been reported even after variceal bleeding[34]. The efficacy of α-receptor blockade has not been evaluated in CF because of the adverse effects of α-blockers on airway reactivity.

There are no studies that have assessed prospectively the indications and actual benefits of different therapeutic interventions and their optimal timing for CF patients with portal hypertension and hypersplenism.

Elective surgical portosystemic shunt was successfully performed in a small series of patients with preserved liver function and without severe pulmonary disease, allowing prolonged postoperative survival up to 15 years[32]. Splenectomy has been performed in association with splenorenal shunt in CF patients with variceal bleeding and massive splenomegaly[67].

Transjugular portosystemic shunt (TIPS) has been employed as a short-term method for portal decompression in patients waiting for liver transplantation[68]. Finally partial splenectomy with conservation of the upper pole of the spleen has been performed in patients with massive splenic enlargement[69]. The efficacy of all these procedures should be further evaluated.

Liver transplantation

Liver transplantation is an effective therapeutic option for CF patients with end-stage liver disease. The 1-year-survival rate after transplantation in CF patients is approximately 80%[13], with beneficial effects on lung function, nutritional status, body composition and quality of life in most cases[39].

Despite the fact that liver transplantation is increasingly performed in CF as a result of improved survival, evidence-based specific guidelines regarding indication and timing are not yet available. As previously mentioned, liver failure is a late event; however, end-stage liver disease may have a major impact on CF. Some authors[70,71] believe that liver transplantation should be considered only in patients with clear hepatocellular failure, in the assumption that, if hepatocellular function is good, survival is dependent almost exclusively on the progression of respiratory disease[34]. Others are more prone to perform transplantation earlier, before signs of hepatocellular failure are prominent, in the assumption that the risk of premature death is increased in patients with cirrhosis and portal hypertension[33]. Recently a pool among European CF and transplant centres was carried out in order to obtain information on current practice and outcome for liver transplant in CF patients in Europe. In the majority of cases the decision to transplant was based on the contemporary presence of various factors and generally before development of end-stage liver disease[43]. Specific clinical scores that consider not only features of portal hypertension, hepatocellular function and hypersplenism, but also nutritional and pulmonary status, have been devised to evaluate the need and timing for liver transplantation in CF patients, but the tests need to be further evaluated[72,73].

CONCLUSIONS

In conclusion, liver disease is a relatively frequent and early complication of CF. The pathogenesis appears to be multifactorial with variable contribution of environmental and genetic determinants. With improved life expectancy of CF patients the impact of liver disease on quality of life and survival is going to increase in future years. UDCA is to date the only accepted treatment for liver disease in CF.

References

1. Rowe SM, Miller S, Sorscher EJ. Cystic fibrosis. N Engl J Med. 2005;352:1992–2001.
2. Davis PB. Cystic fibrosis since 1938. Am J Respir Crit Care Med. 2006;173;475–82.
3. Riordan JR, Rommens JM, Kerem B et al. Identification of the cystic fibrosis gene: cloning and characterization of complementary DNA. Science. 1989;245:1066–73.
4. Mehta A. CFTR: more than just a chloride channel. Pediatr Pulmonol. 2005;39:292–8.
5. www.genet.sickkids.on.ca/cftr/app
6. Kerem E, Kerem B. Genotype–phenotype correlations in cystic fibrosis. Pediatr Pulmonol. 1996;22:387–95.
7. Kristidis P, Bozon D, Coorey M et al. Genetic determination of exocrine pancreatic function in cystic fibrosis. Am J Hum Genet. 1992;50.1178–84.
8. Knowles MR, Durie PR. What is cystic fibrosis? N Engl J Med. 2002;347:439–42.
9. Colombo C, Battezzati PM. Liver involvement in cystic fibrosis: primary organ damage or innocent bystander? J Hepatol. 2004;41:1041–44.
10. Shapira R, Hadzic R, Francavilla R, Koukulis G, Price GF, Mieli-Vergani G. Retrospective review of cystic fibrosis presenting as infantile liver disease. Arch Dis Child. 1999;81:125–8.
11. Lindblad A, Glaumann H, Strandvik B. Natural history of liver disease in cystic fibrosis. Hepatology. 1999;30:1151–8.

12. Chen AH, Innis SM, Davidson GF, Jill James S. Phosphatidylcholine and lysophosphatidyl-choline excretion is increased in children with cystic fibrosis and is associated with plasma homocysteine, *S*-adenosylhomocysteine, and *S*-adenosylmethionine. Am J Clin Nutr. 2005;81:686–91.
13. Feranchak AP, Sokol RJ. Cholangiocyte biology and cystic fibrosis liver disease. Semin Liver Dis. 2001;21:471–88.
14. Pinto HC, Carniero de Moura M, Day CP. Non-alcoholic steatohepatitis from cell biology to clinical practice. J Hepatol. 2006;44:197–208.
15. Gaskin KJ, Waters DLM, Howman-Giles R et al. Liver disease and common bile-duct stenosis in cystic fibrosis. N Engl J Med. 1988;318:340–6.
16. Standvik B, Hjelte L, Gabrielsson N, Glaumann H. Sclerosing cholangitis in cystic fibrosis. Scand J Gastroenterol Suppl. 1988;143:121–4.
17. Durieu I, Pellet O, Simonot L et al. Sclerosing cholangitis in adults with cystic fibrosis: a magnetic resonance cholangiographic prospective study. J Hepatol. 1999;30:1052–6.
18. Sokol RJ, Durie PR. Recommendations for management of liver and biliary tract disease in cystic fibrosis. Cystic Fibrosis Foundation Hepatobiliary Disease Consensus Group. J Pediatr Gastoenterol Nutr. 1999:28(Suppl.1):1–13.
19. Cystic Fibrosis Foundation, Patient Registry. 2003 Annual Report to the Center Directors. Bethesda, Maryland; Cystic Fibrosis Foundation, 2004.
20. Cohn JA, Strong TV, Picciotto MR, Nairn AC, Collins FS, Fitz JG. Localization of the cystic fibrosis transmembrane conductance regulator in human bile duct epithelial cells. Gastroenterology. 1993;105:1857–64.
21. Lindblad A, Hultcrantz R, Strandvik B. Bile duct destruction and collagen deposition: a prominent ultrastructural feature of the liver in cystic fibrosis. Hepatology. 1992;16:372–81.
22. Bhaskar KR, Turner BS, Grubman SA, Jefferson DM, LaMont JT. Dysregulation of proteoglycan production by intrahepatic biliary epithelial cells bearing defective (delta-f508) cystic fibrosis transmembrane regulator. Hepatology. 1998;27:54–61.
23. Oppenheimer EH, Esterly JR. Hepatic changes in young infants with cystic fibrosis: possible relation to focal biliary cirrhosis. J Pediatr. 1975;86:683–9.
24. Sinaasappel M. Hepatobiliary pathology in patients with cystic fibrosis. Acta Paediatr Scand Suppl. 1989;363:45–51.
25. Colombo C, Apostolo MG, Ferrari M et al. Analysis of risk factors for the development of liver disease associated with cystic fibrosis. J Pediatr. 1994;124:393–9.
26. Wilschanski M, Rivlin J, Cohen S et al. Clinical and genetic risk factors for CF-related liver disease. Pediatrics. 1999;103:52–7.
27. Castaldo G, Fuccio A, Salvatore D et al. Liver expression in cystic fibrosis could be modulated by genetic factors different from the cystic cibrosis transmembrane regulator genotype. Am J Med Genet. 2001;98:294–7.
28. Knowles MR, Drumm ML, Konstan MW et al., Gene Modifier Study Group. Genetic modifiers of lung disease in cystic fibrosis. N Engl J Med. 2005;353:1443–53.
29. Friedman KJ, Ling SC, Lange EM et al. Genetic modifiers of severe liver disease in cystic fibrosis. Pediatr Pulmon. 2005;40(Suppl. 28):247 (Abstract).
30. Vawter GF, Shwachman H. Cystic fibrosis in adults: an autopsy study. Pathol Ann. 1979;14: 357–82.
31. Colombo C, Battezzati PM, Podda M. Hepatobiliary disease in cystic fibrosis. Semin Liver Dis. 1994;14:259–69.
32. Debray D, Lykavieris P, Gauthier F et al. Outcome of cystic fibrosis-associated liver cirrhosis: management of portal hypertension. J Hepatol. 1999;31:77–83.
33. Feigelson J, Anagnostopoulos C, Poquet M, Pecau Y, Munck A, Navarro J. Liver cirrhosis in cystic fibrosis. Therapeutic implications and long-term follow-up. Arch Dis Child. 1993;68:653–7.
34. Gooding I, Dondis V, Gyi K M, Hodson M, Westaby D. Variceal hemorrhage and cystic fibrosis: outcomes and implications for liver transplantation. Liver Transplant. 2005;12: 1522–6.
35. Ling SC, Wilkinson JD, Hollman AS, McColl J, Evans TJ, Paton JY. The evolution of liver disease in cystic fibrosis. Arch Dis Child. 1999;81:129–32.
36. Colombo C, Battezzati PM, Crosignani A et al. Liver disease in cystic fibrosis: a prospective study on incidence, risk factors and outcome. Hepatology. 2002;36:1374–82.

37. Pencharz PB, Durie PR. Pathogenesis of malnutrition in cystic fibrosis, and its treatment. Clin Nutr. 2000;19:387–94.
38. Marshall BC, Butler SM, Stoddard M, Moran AM, Liou TG, Morgan WJ. Epidemiology of cystic fibrosis-related diabetes. J Pediatr. 2005;146:681–7.
39. Colombo C, Costantini D, Rocchi A et al. Effects of liver transplantation on the nutritional status of patients with cystic fibrosis. Transplant Int. 2005;18:246–55.
40. Lamireau T, Monnereau S, Martin S et al. Epidemiology of liver disease in cystic fibrosis: a longitudinal study. J Hepatol. 2004;41:920–5.
41. Corbett K, Kelleher S, Rowland M et al. Cystic fibrosis-associated liver disease: a population-based study. J Pediatr. 2004;145:327–32.
42. Scott-Jupp R, Lama M, Tanner MS. Prevalence of liver disease in cystic fibrosis. Arch Dis Child. 1991;66:698–701.
43. Melzi ML, Kelly D, Colombo C et al. Liver transplant in cystic fibrosis: a pool among European centers. A study from the European Liver Transplant Registry. Transplant Int. 2006;19:726–31.
44. Maurage C, Lenaerts C, Weber AM et al. Meconium ileus and its equivalent as a risk factor for the development of cirrhosis: an autopsy study in cystic fibrosis. J Pediatr Gastroenterol Nutr. 1989;9:17–20.
45. Farrell MP. Guidelines for implementation of cystic fibrosis newborn screening programs: Cystic Fibrosis Foundation Workshop Report. Pediatrics. 2007;119:e495–518.
46. Potter CJ, Fishbein M, Hammond S et al. Can the histologic changes of cystic fibrosis-associated hepatobiliary disease be predicted by clinical criteria? J Pediatr Gastroenterol Nutr. 1997;25:32–6.
47. Patriquin H, Lenaerts C, Smith L et al. Liver disease in children with cystic fibrosis: US and biochemical comparison in 195 patients. Radiology. 1999;211:229–32.
48. Lenaerts C, Lapierre C, Patriquin H et al. Surveillance for cystic fibrosis-associated hepatobiliary disease: early ultrasound changes and predisposing factors. J Pediatr. 2003; 143:343–50.
49. Williams SGJ, Evanson JE, Barrett N, Hodson ME, Boultbee JE, Westaby D. An ultrasound scoring system for the diagnosis of liver disease in cystic fibrosis. J Hepatol. 1995;22:513–21.
50. Stewart L. The role of abdominal ultrasound in the diagnosis, staging and managment of cystic fibrosis liver disease. J Roy Soc Med. 2005;98:17–27.
51. O'Connor PJ, Southern KW, Bowler IM. The role of hepatobiliary scintigraphy in cystic fibrosis. Hepatology. 1996;23:281–7.
52. Foster JA, Ramsden WH, Conway SP, Taylor JM, Etherington C. The role of IDA scintigraphy in the follow-up of liver disease in patients with cystic fibrosis. Nucl Med Commun. 2002;23:673–81.
53. Colombo C, Castellani MR, Balistreri WF, Seregni E, Assaisso ML, Giunta A. Scintigraphic documentation of an improvement in hepatobiliary excretory function after treatment with ursodeoxycholic acid in patients with cystic fibrosis and associated liver disease. Hepatology. 1992;15:677–84.
54. McKeon D, Day A, Parmar J, Alexander G, Bilton D. Hepatocellular carcinoma in association with cirrhosis in a patient with cystic fibrosis. J Cyst Fibros. 2004;3:193–5.
55. Yang Y, Raper SE, Cohn JA, Engelhardt JF, Wilson JM. An approach for treating the hepatobiliary disease of cystic fibrosis by somatic gene transfer. Proc Natl Acad Sci USA. 1993;90:4601–5.
56. Roman RM, Feranchak AP, Salter KD et al. Endogenous ATP regulates Cl^- secretion in cultured human and rat biliary epithelial cells. Am J Physiol. 1999;276:G1391–400.
57. Spirli C, Fiorotto R, Song L, Santos-Sacchi J, Okolicsany L, Maser S. Glibenclamide stimulates fluid secretion in rodent cholangiocytes through a cystic fibrosis transmembrane conductance regulator-independent mechanism. Gastroenterology. 2005;129:220–33.
58. Paumgartner G, Beuers U. Ursodeoxycholic acid in cholestatic liver disease: mechanisms of action and therapeutic use revisited. Hepatology. 2002;36:525–31.
59. Chinet T, Fouassier L, Dray-Charier N et al. Regulation of electrogenic anion secretion in normal and cystic fibrosis gallbladder mucosa. Hepatology. 1999;29:5–13.
60. Colombo C, Crosignani A, Assaisso ML et al. Ursodeoxycholic acid therapy in cystic fibrosis associated liver disease: a dose–response study. Hepatology. 1992;16:924–30.

61. Colombo C, Battezzati PM, Podda M, Bettinardi N, Giunta A and the Italian Group for the Study of Ursodeoxycholic Acid in Cystic Fibrosis. Ursodeoxycholic acid for liver disease associated with cystic fibrosis: a double-blind multicenter trial. Hepatology. 1996; 23:1484–90.
62. Lindblad A, Glaumann H, Strandvik B. A two-year prospective study of the effect of ursodeoxycholic acid on urinary bile acid excretion and liver morphology in cystic fibrosis. Hepatology. 1998;23:166–74.
63. Lepage G, Paradis K, Lacaille F et al. Ursodeoxycholic acid improves the hepatic metabolism of essential fatty acids and retinol in chidren with cystic fibrosis. J Pediatr. 1997;130:52–8.
64. Freedman SD, Blanco PG, Zaman MM et al. Association of cystic fibrosis with abnormalities in fatty acid metabolism. N Engl J Med. 2004;350:560–9.
65. Nousia-Arvanitakis S, Futoulaki M, Economou H et al. Long-term prospective study of the effect of ursodeoxycholic acid on cystic fibrosis-related liver disease. J Clin Gastroenterol. 2001;32:324–8.
66. Cheng K, Ashby D, Smyth R. Ursodeoxycholic acid for cystic fibrosis-related liver disease (Cochrane Review). Cochrane Library, Issue 2. Oxford: Update Software 6-12, 2000.
67. Robberecht E, Van Biervliet S, Vanrentergem K, Kerremans I. Outcome of total splenectomy with portosystemic shunt for massive splenomegaly and variceal bleeding in cystic fibrosis. J Pediatr Surg. 2006;41:1561–5.
68. Pozler O, Krajina A, Vanicek H et al. Transjugular intrahepatic portosystemic shunt in five children with cystic fibrosis: long-term results. Hepatogastroenterology. 2003;50:1111–14.
69. Zach MS, Thalhammer GH, Eber E. Partial splenectomy in CF patients with hypersplenism. Arch Dis Child. 2003;88:649.
70. Mack DR, Traystman MD, Colombo JL et al. Clinical denouement and mutation analysis of patients with cystic fibrosis undergoing liver transplantation for biliary cirrhosis. J Pediatr. 1995;127:881.
71. Sharp HL. Cystic fibrosis liver disease and transplantation. J Pediatr. 1995;127:944.
72. Noble-Jamieson G, Barnes N, Jamieson N, Friend P, Cain R. Liver transplantation for hepatic cirrhosis in cystic fibrosis. J Roy Soc Med. 1996;89:31–7.
73. Milkiewicz P, Skiba G, Kelly D et al. Transplantation for cystic fibrosis: outcome following early liver transplantation. J Gastroenterol Hepatol. 2002;17:208.

Section III
Liver diseases (Part 2)

Chair: DW COX and GA KULLAK-UBLICK

13
State-of-the-Art Lecture III: Genetics of cholestasis

P. L. M. JANSEN

Cholestasis results from a complete or partial cessation of bile flow. As a consequence solutes that normally are secreted into bile accumulate in liver, blood and extrahepatic tissues. Among the various solutes in bile, bile salts are thought to be metabolically the most active and toxic elements. In bile their toxicity is neutralized by phospholipids and cholesterol, but upon accumulation in the hepatic parenchyma they are harmful by their apoptotic, fibrogenic, proinflammatory and oncogenic actions. Since the liver is specialized in the uptake and concentration of bile salts it is not surprising that liver and bile ducts are the main victims of bile salt toxicity during cholestasis. Although cholestasis is a common denominator, the various cholestatic liver diseases differ in their phenotype. Primary biliary cirrhosis and primary sclerosing cholangitis can be considered as acquired polygenetic diseases; the various subtypes of progressive familial intrahepatic cholestasis as inborn monogenetic diseases.

In primary biliary cirrhosis (PBC) the primary lesion is an inflammatory infiltrate of T-lymphocyte infiltrates slowly destroying the bile ductules. The vast majority of patients are females. In primary sclerosing cholangitis (PSC) the primary lesion is portal and periportal inflammation and stricturing fibrosis of the larger extrahepatic or intrahepatic bile ducts. Cholangiocarcinoma is a major complication, and one of the big problems in the management of this disease. This is a deadly complication without many therapeutic options.

Drug-induced cholestatic liver disease is rather common. It is usually reversible upon discontinuation of the medication. In rare cases drug-induced cholestatic liver disease proceeds to a vanishing bile duct syndrome in which the bile ducts proceed from a phase of ductopenia to a situation of a dramatic paucity of bile ducts with chronic cholestasis and eventually biliary fibrosis and cirrhosis.

Although the aetiopathogenesis of these diseases is multifactorial, genes play a role. In PBC an interaction between genes and the environment is postulated. Epidemiological studies show that PBC preferentially occurs in industrialized areas or near waste dumps, and the search is for a toxic agent that has entered the water supply or food chain, and that may be secreted and concentrated in bile[1–4]. Metabolism of this putative agent in the liver may be the genetic and toxifying component in the aetiology of this disease. An alternative view is that a bacterium or retrovirus is somehow involved in the aetiopathogenesis of the disease[2,5]. Family and twin studies confirm that, in addition to these environmental influences, there is a strong genetic component[6].

In PSC there is a striking relation with inflammatory bowel disease, in particular colitis. It is tempting to speculate that this may provide a lead to a possible aetiopathogenesis involving the colon. There are no animal models that directly link induced colitis to PSC. The Mdr2$^{-/-}$ knockout mouse has pericholangitis and fibrosis with some resemblance to PSC[7]; however, these mice have no colitis; therefore the pathogenesis remains puzzling and is far from resolved. Associations with a certain polymorphism of PXR/SXR suggest that this nuclear hormone receptor may act as a disease-modifier gene[8].

Monogenetic diseases provide insight in the importance of a particular gene for liver function and bile formation. For some genes a clear relation exists between severity of disease and a gene mutation. The severest disease expression is seen in situations wherein either the disease-related protein is not produced or is mutated in one of its essential functions such as the Walker domains or ATP-binding site of ABC transporters. Knockout animal models allow detailed studies of these diseases. Phenotypes in animal models may differ from that in humans. In many instances this is due to a less toxic bile salt pattern in mice than in humans. Table 1 gives an overview of relevant disease models and Table 2 shows corresponding human diseases.

The gene defect underlying progressive familial intrahepatic cholestasis, PFIC type 1, was discovered by classical haplotype sharing analysis[9]. Clinically, this disease had already been characterized long before the gene was found. This disease was known as Byler's disease, a cholestatic disease that occurs among the Pennsylvanian Amish community. It is now recognized that *FIC1* gene mutations also occur outside the Amish community. The phenotype of the animal model for this disease is the Atp8b1 G308V/G308V mutant mice. This mouse carries a mutation orthologous to that present in patients with Byler's disease[10]. However, in contrast to patients with Byler's disease, these mice are not cholestatic, and in fact have an increased bile flow. Despite this difference the mutant mice have allowed a detailed study on the function of FIC1 and the consequence of the lack thereof.

FIC1 is an aminophospholid flippase with an outward to inward activity, flipping phosphatidylserine from the outer to the inner leaflet of membranes. Its expression is highest in cholangiocytes and ileum mucosa cells[11]. It is also expressed in the canalicular membrane of hepatocytes. FIC1 has been proposed as a regulator of FXR activity[12,13] but current attention is given to FIC1 as a membrane-stabilizing factor[14]. This function sheds light on the low serum gamma-glutamyltransferase (γ-GT) activity in patients with Byler's disease. Not only is the serum γ-GT not elevated in these patients, but the serum γ-GT is often abnormally low. In the liver γ-GT is an anchored protein on the luminal side of canaliculi. In mutant mice, and in patients with FIC1 deficiency, the canalicular membrane becomes unstable and γ-GT, together with other anchored membrane proteins, is eluted in bile, thereby reducing γ-GT activity in liver and serum. Despite this finding the link between FIC1 deficiency and

Table 1 Mouse models for cholestatic diseases

			Gene knockout or gene mutation mouse models	
	Gene	*Protein*	*Control conditions*	*Stressed conditions (BDL, cholate/drug exposure)*
Constitutive	ABCB11	BSEP	None	Mild cholestasis[18]
	ABCB4	Mdr2/MDR3	Mild fibrosis	Severe fibrosis, inflammation, CC/HCC[31,32]
	ABCG5/G8	ABC G5/G8	None	Reduced cholesterol secretion[33]
	ABCC2	MRP2	Jaundice	Altered drug metabolism[34]
	ATP8B1*	FIC1	Increased protein in bile	?[10,14]
Regulation	FXR	FXR	Mild steatosis	CC/HCC[35–38]
	PXR/SXR	PXR	Colitis	BDL: more necrosis[39–42]
	CAR	CAR	None	BDL: more necrosis[41–44]
Adaptation	ABCC3	MRP3	None	BDL: reduced hyperbilirubinaemia[45,46]
	ABCC4	MRP4	None	BDL: more necrosis[47,48]
	ABCC1	MRP1	None	Sensitivity towards xenotoxicants[49]
	ABCG2	BCRP	Phototoxicity	Sensitivity towards xenotoxicants[50–52]

CC, cholangiocarcinoma; HCC, hepatocellular carcinoma; BDL, bile duct ligation.

Table 2 Human genetic intrahepatic cholestatic diseases

Disease: *Gene*	*PFIC type 1* *ATP8B1/FIC1*	*PFIC type 2* *ABCB11/BSEP*	*PFIC type 3* *ABCB4/MDR3*
Histology	Bland cholestasis	Neonatal hepatitis, cholestasis, portal fibrosis	Bile duct proliferation, portal fibrosis
Immune histochemistry	FIC1 absent	BSEP absent	MDR 3 absent
Serum γ-GT activity	Low γ-GT	Low γ-GT	High γ-GT
Clinical phenotype	Intermittent cholestasis, severe pruritus; age of onset: 0–1 year	Permanent cholestasis, severe pruritus; age of onset: 0–1 year	Permanent cholestasis, severe pruritus; age of onset: 1–5 year
Therapy	Biliary diversion, transplantation	Biliary diversion, transplantation	Ursodeoxycholic acid, transplantation
Benign manifestations	BRIC type 1, ICP	BRIC type 2, ICP	Intrahepatic gallstones, sludge, ICP
Complications		HCC and CC <5 years of age	Vanishing bile duct syndrome

BRIC, benign recurrent intrahepatic cholestasis; HCC, hepatocellular carcinoma; ICP, intrahepatic cholestasis of pregnancy.

cholestasis has not been elucidated. The phenotypic difference between mice and humans with the same FIC1 mutation may be the consequence of modifier genes that prevent cholestasis in mice, or it may relate to the old observation that cholestasis in patients with benign recurrent intrahepatic cholestasis (a more benign variant of Byler's of PFIC1 disease) is related to toxic bile salt species[15]. These may be produced in humans but not in mice.

By identifying mutations in the *ABCB11* gene as the underlying cause of progressive familial intrahepatic cholestasis type 2 in humans it became clear that the protein product, BSEP (ABCB11) or bile acid export pump, is the main bile acid transporter in the canalicular membrane[16,17]. Also here the human disease, PFIC type 2, presents with considerably more cholestasis than the mouse model, the *Abcb11*$^{-/-}$ knockout mice[18]. Had it only been for knockout mice the importance of BSEP as the main bile salt transporter may not have been appreciated.

MDR3 (ABCB4) is a canalicular transporter of phospholipids, particularly of phosphatidylcholine. A number of MDR3 gene mutations have been identified in patients with progressive familial intrahepatic cholestasis type 3. Here the *mdr2*$^{-/-}$ knockout mouse provided a disease model with most elements of the human disease, albeit that the phenotype became fully manifest only after feeding the mice the 'human' bile salts cholate and chenodeoxycholate. The human disease, progressive familial intrahepatic cholestasis type 3, was discovered shortly after it was realized that phospholipids in bile serve to neutralize the cytotoxicity of bile acids, and that bile duct damage and proliferation are key elements in diseases wherein MDR3 is lacking. Thus, in this case, knockout mice have led the way to the human disease.

All three subtypes of progressive familial intrahepatic cholestasis have more benign counterparts. Genetically the more benign phenotype can be traced to either heterozygous mutations or to homozygous mutations in functionally more silent regions of the transport protein. Intrahepatic cholestasis of pregnancy can be the consequence of mutations in any of the three genes: *ATP8B1*, *ABCB11* or *ABCB4*[19–24]. In these patients cholestasis occurs in the third trimester of pregnancy. Some of these patients also show some signs of cholestasis, such as pruritus, when using oral anticontraceptive medication. Benign recurrent intrahepatic cholestasis is associated with mutations of either the *ATP8B1* or *ABCB11* genes[25,26].

Do any of these cholestatic syndromes bring us any further in understanding the more complex entities such as PBC or PSC? Although the mdr 2$^{-/-}$ knockout mouse has been called a disease model for either of these diseases, it differs from these diseases in many respects. Yet the recognition that lack of phosphatidylcholine in bile leads to bile salt toxicity, that affects both bile ducts and hepatocytes, has been a great step forward in bile (patho)physiology. Both in PBC and PSC part of the evil could come from the bile in the bile ductules. Leakage of bile constituents from the bile ductules into the surrounding parenchyma could aggravate the periportal inflammation and fibrosis that characterizes both diseases. One may speculate that in PBC this leakage may be the result of damage of bile duct epithelium, while in PSC leaky tight junctions may be more likely[27].

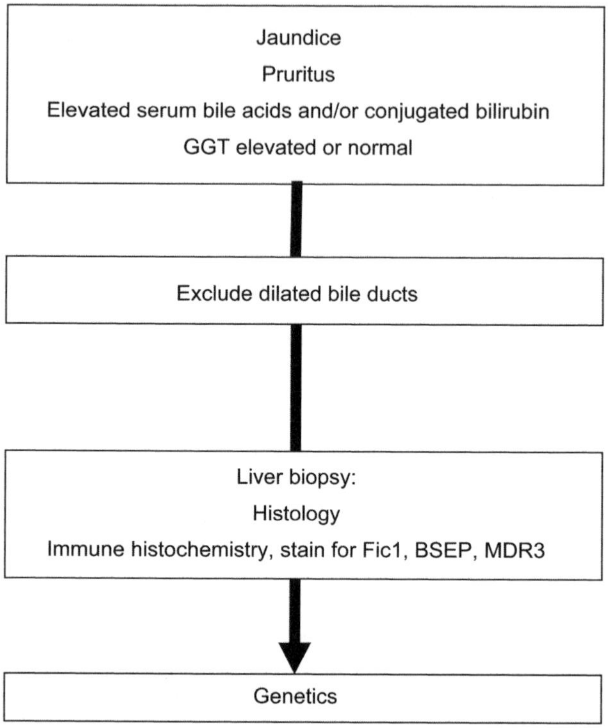

Figure 1 Clinical evaluation of patients with suspected progressive familial intrahepatic cholestasis

Other relevant transporters in the hepatocyte for which gene polymorphisms and mutations have been described are MDR1, a xenobiotic transporter; ABC G5/G8, a cholesterol transporter; MRP2, the major transporter for a great number of organic anions including bilirubin conjugates; and BCRP, a transporter of xenobiotics that also may play a cytoprotective function in hepatic progenitor cells[28–30]. Drugs, non-bile acid endogenous metabolites or environmental waste products are substrates for these transporters. Defects in these proteins could potentially lead to accumulation of toxic agents or defects of regeneration. Thus any of these transporters may play a role in liver disease.

For the clinician the difference between the various form of progressive familial intrahepatic cholestasis is not always that clear. The algorithm in Figure 1 may help to make the diagnosis. If intrahepatic cholestasis is suspected, standard liver histology is a first but important step. Bland cholestasis is compatible with PFIC type 1. PFIC type 2 often presents as neonatal hepatitis and in PFIC type 3 there is portal inflammation and bile duct proliferation. Immunohistochemistry, using antibodies against any of the relevant transport proteins, is a great tool that assists in the diagnosis. Genetic analysis is a final step to arrive at the correct diagnosis.

References

1. Arbour L, Rupps R, Field L et al. Characteristics of primary biliary cirrhosis in British Columbia's First Nations population. Can J Gastroenterol. 2005;19:305–10.
2. Selmi C, Gershwin ME. Bacteria and human autoimmunity: the case of primary biliary cirrhosis. Curr Opin Rheumatol. 2004;16:406–10.
3. Jones DE, Watt FE, Metcalf JV, Bassendine MF, James OF. Familial primary biliary cirrhosis reassessed: a geographically-based population study. J Hepatol. 1999;30:402–7.
4. Prince MI, Chetwynd A, Diggle P, Jarner M, Metcalf JV, James OF. The geographical distribution of primary biliary cirrhosis in a well-defined cohort. Hepatology. 2001;34: 1083–8.
5. Mason A, Xu L, Neuberger J. Proof of principal studies to assess the role of the human betaretrovirus in patients with primary biliary cirrhosis. Am J Gastroenterol. 2004;99: 2499–500.
6. Selmi C, Mayo MJ, Bach N et al. Primary biliary cirrhosis in monozygotic and dizygotic twins: genetics, epigenetics, and environment. Gastroenterology. 2004;127:485–92.
7. Popov Y, Patsenker E, Fickert P, Trauner M, Schuppan D. Mdr2 (Abcb4)$^{-/-}$ mice spontaneously develop severe biliary fibrosis via massive dysregulation of pro- and antifibrogenic genes. J Hepatol. 2005;43:1045–54.
8. Karlsen TH, Lie BA, Frey FK et al. Polymorphisms in the steroid and xenobiotic receptor gene influence survival in primary sclerosing cholangitis. Gastroenterology. 2006;131:781–7.
9. Bull LN, Carlton VE, Stricker NL et al. Genetic and morphological findings in progressive familial intrahepatic cholestasis (Byler disease [PFIC-1] and Byler syndrome): evidence for heterogeneity. Hepatology. 1997;26:155–64.
10. Pawlikowska L, Groen A, Eppens EF et al. A mouse genetic model for familial cholestasis caused by ATP8B1 mutations reveals perturbed bile salt homeostasis but no impairment in bile secretion. Hum Mol Genet. 2004;13:881–92.
11. Eppen EF, van Mil SW, De Vree JM et al. FIC1, the protein affected in two forms of hereditary cholestasis, is localized in the cholangiocyte and the canalicular membrane of the hepatocyte. J Hepatol. 2001;35:436–43.
12. Chen F, Ananthanarayanan M, Emre S et al. Progressive familial intrahepatic cholestasis, type 1, is associated with decreased farnesoid X receptor activity. Gastroenterology. 2004; 126:756–64.
13. Alvarez L, Jara P, Sanchez-Sabate E et al. Reduced hepatic expression of farnesoid X receptor in hereditary cholestasis associated to mutation in ATP8B1. Hum Mol Genet. 2004;13:2451–60.
14. Paulusma CC, Groen A, Kunne C et al. Atp8b1 deficiency in mice reduces resistance of the canalicular membrane to hydrophobic bile salts and impairs bile salt transport. Hepatology. 2006;44:195–204.
15. Bijleveld CM, Vonk RJ, Kuipers F, Havinga R, Fernandes J. Benign recurrent intrahepatic cholestasis: a long-term follow-up study of two patients. Hepatology. 1989;9:532–7.
16. Strautnieks SS, Bull LN, Knisely AS et al. A gene encoding a liver-specific ABC transporter is mutated in progressive familial intrahepatic cholestasis. Nat Genet. 1998;20:233–8.
17. Jansen PL, Strautnieks SS, Jacquemin E et al. Hepatocanalicular bile salt export pump deficiency in patients with progressive familial intrahepatic cholestasis. Gastroenterology. 1999;117:1370–9.
18. Wang R, Salem M, Yousef IM et al. Targeted inactivation of sister of P-glycoprotein gene (spgp) in mice results in nonprogressive but persistent intrahepatic cholestasis. Proc Natl Acad Sci USA. 2001;98:2011–16.
19. de Pagter AG, Berge Henegouwen GP, Bokkel Huinink JA, Brandt KH. Familial benign recurrent intrahepatic cholestasis. Interrelation with intrahepatic cholestasis of pregnancy and from oral contraceptives? Gastroenterology. 1976;71:202–7.
20. Dixon PH, Weerasekera N, Linton KJ et al. Heterozygous MDR3 missense mutation associated with intrahepatic cholestasis of pregnancy: evidence for a defect in protein trafficking. Hum Mol Genet. 2000;9:1209–17.
21. Gendrot C, Bacq Y, Brechot MC, Lansac J, Andres C. A second heterozygous MDR3 nonsense mutation associated with intrahepatic cholestasis of pregnancy. J Med Genet. 2003;40:e32.

22. Jacquemin E, Cresteil D, Manouvrier S, Boute O, Hadchouel M. Heterozygous non-sense mutation of the MDR3 gene in familial intrahepatic cholestasis of pregnancy. Lancet. 1999; 353:210–11.
23. Pauli-Magnus C, Lang T, Meier Y et al. Sequence analysis of bile salt export pump (ABCB11) and multidrug resistance p-glycoprotein 3 (ABCB4, MDR3) in patients with intrahepatic cholestasis of pregnancy. Pharmacogenetics. 2004;14:91–102.
24. De Vree JM, Jacquemin E, Sturm E et al. Mutations in the MDR3 gene cause progressive familial intrahepatic cholestasis. Proc Natl Acad Sci USA. 1998;95:282–7.
25. Bull LN, van Eijk MJ, Pawlikowska L et al. A gene encoding a P-type ATPase mutated in two forms of hereditary cholestasis. Nat Genet. 1998;18:219–24.
26. van Mil SW, van der Woerd WL, van der BG et al. Benign recurrent intrahepatic cholestasis type 2 is caused by mutations in ABCB11. Gastroenterology. 2004;127:379–84.
27. Fickert P, Fuchsbichler A, Wagner M et al. Regurgitation of bile acids from leaky bile ducts causes sclerosing cholangitis in Mdr2 (Abcb4) knockout mice. Gastroenterology. 2004;127: 261–74.
28. Ros JE, Libbrecht L, Geuken M, Jansen PL, Roskams TA. High expression of MDR1, MRP1, and MRP3 in the hepatic progenitor cell compartment and hepatocytes in severe human liver disease. J Pathol. 2003;200:553–60.
29. Ros JE, Roskams TA, Geuken M et al. ATP binding cassette transporter gene expression in rat liver progenitor cells. Gut. 2003;52:1060–7.
30. Van der Bogt S, Libbrecht L, Katoonizadeh A et al. Breast cancer resistance protein (BCRP/ABCG2) is expressed by progenitor cells/reactive ductules and hepatocytes and its expression pattern is influenced by disease etiology and species type: possible functional consequences. J Histochem Cytochem. 2006;54:1051–9.
31. Smit JJ, Mol CA, van Deemter L, Wagenaar E, Schinkel AH, Borst P. Characterization of the promoter region of the human MDR3 P-glycoprotein gene. Biochim Biophys Acta. 1995;1261:44–56.
32. Mauad TH, van Nieuwkerk CM, Dingemans KP et al. Mice with homozygous disruption of the mdr2 P-glycoprotein gene. A novel animal model for studies of nonsuppurative inflammatory cholangitis and hepatocarcinogenesis. Am J Pathol. 1994;145:1237–45.
33. Klett EL, Lu K, Kosters A et al. A mouse model of sitosterolemia: absence of Abcg8/ sterolin-2 results in failure to secrete biliary cholesterol. BMC Med. 2004;2:5.
34. Paulusma CC, Bosma PJ, Zaman GJ et al. Congenital jaundice in rats with a mutation in a multidrug resistance-associated protein gene. Science. 1996;271:1126–8.
35. Sinal CJ, Tohkin M, Miyata M, Ward JM, Lambert G, Gonzalez FJ. Targeted disruption of the nuclear receptor FXR/BAR impairs bile acid and lipid homeostasis. Cell. 2000;102:731–44.
36. Kim I, Morimura K, Shah Y, Yang Q, Ward JM, Gonzalez FJ. Spontaneous hepatocarcinogenesis in farnesoid X receptor-null mice. Carcinogenesis. 2006; December 20 [Epub ahead of print].
37. Marschall HU, Wagner M, Bodin K et al. Fxr(–/–) mice adapt to biliary obstruction by enhanced phase I detoxification and renal elimination of bile acids. J Lipid Res. 2006;47: 582–92.
38. Wagner M, Fickert P, Zollner G et al. Role of farnesoid X receptor in determining hepatic ABC transporter expression and liver injury in bile duct-ligated mice. Gastroenterology. 2003;125:825–38.
39. Xie W, Radominska-Pandya A, Shi Y et al. An essential role for nuclear receptors SXR/ PXR in detoxification of cholestatic bile acids. Proc Natl Acad Sci USA. 2001;98:3375–80.
40. Uppal H, Toma D, Saini SP, Ren S, Jones TJ, Xie W. Combined loss of orphan receptors PXR and CAR heightens sensitivity to toxic bile acids in mice. Hepatology. 2004;41:168–76.
41. Wagner M, Halilbasic E, Marschall HU et al. CAR and PXR agonists stimulate hepatic bile acid and bilirubin detoxification and elimination pathways in mice. Hepatology. 2005; 42:420–30.
42. Stedman CA, Liddle C, Coulter SA et al. Nuclear receptors constitutive androstane receptor and pregnane X receptor ameliorate cholestatic liver injury. Proc Natl Acad Sci USA. 2005;102:2063–8.
43. Huang W, Zhang J, Chua SS et al. Induction of bilirubin clearance by the constitutive androstane receptor (CAR). Proc Natl Acad Sci USA. 2003;100:4156–61.

44. Huang W, Zhang J, Washington M et al. Xenobiotic stress induces hepatomegaly and liver tumors via the nuclear receptor constitutive androstane receptor. Mol Endocrinol. 2005;19: 1646–53.
45. Belinsky MG, Dawson PA, Shchaveleva I et al. Analysis of the *in vivo* functions of Mrp3. Mol Pharmacol. 2005;68:160–8.
46. Zelcer N, van de WK, de WR et al. Mice lacking Mrp3 (Abcc3) have normal bile salt transport, but altered hepatic transport of endogenous glucuronides. J Hepatol. 2006;44: 768–75.
47. Borst P, de Wolf C, van de Wetering K. Multidrug resistance-associated proteins 3, 4, and 5. Pflugers Arch. 2007;453:661–73.
48. Mennone A, Soroka CJ, Cai SY et al. Mrp4–/– mice have an impaired cytoprotective response in obstructive cholestasis. Hepatology. 2006;43:1013–21.
49. Wijnholds J, Scheffer GL, van der Valk M et al. Multidrug resistance protein 1 protects the oropharyngeal mucosal layer and the testicular tubules against drug-induced damage. J Exp Med. 1998;188:797–808.
50. Jonker JW, Merino G, Musters S et al. The breast cancer resistance protein BCRP (ABCG2) concentrates drugs and carcinogenic xenotoxins into milk. Nat Med. 2005;11: 127–9.
51. Jonker JW, Buitelaar M, Wagenaar E et al. The breast cancer resistance protein protects against a major chlorophyll-derived dietary phototoxin and protoporphyria. Proc Natl Acad Sci USA. 2002;99:15649–54.
52. Merino G, Jonker JW, Wagenaar E, van Herwaarden AE, Schinkel AH. The breast cancer resistance protein (BCRP/ABCG2) affects pharmacokinetics, hepatobiliary excretion, and milk secretion of the antibiotic nitrofurantoin. Mol Pharmacol. 2005;67:1758–64.

14
Hereditary liver diseases: screening strategies

P. T. CLAYTON

INTRODUCTION

Screening for genetic liver diseases in infancy can be selective or population-based. In selective screening infants with evidence of liver disease are tested for a genetic disorder (or more usually a range of genetic disorders), or an infant with a family history of a genetic liver disease is tested for that disease. In population screening all newborns are tested for one or more genetic liver diseases. Population screening programmes are usually based on blood spots that are collected in the first 10 days of life.

Examples of techniques already in widespread use for selective screening are shown in Table 1; they include tests at the metabolite, protein and DNA levels. The great technological advances that have been made (particularly in the area of mass spectrometry) mean that metabolite- and protein-based screening methods can often be taken from a selective screening laboratory to a population screening setting. Whether this is desirable or not will be discussed below. However, first I will describe some of our recent experience with selective and population screening and some of the specific technological advances that are relevant to the discussion of screening strategies.

GC-MS ANALYSIS OF THE PLASMA STEROL PROFILE

Among the tests we now routinely undertake by GC-MS is analysis of the plasma sterol profile. A normal profile shows cholesterol at 2.5–5.5 mM concentration, a number of cholesterol precursors at concentrations <20 μM and phytosterols at concentrations <60 μM. Figure 1 shows a profile in which phytosterols are present at millimolar concentration. We first observed this profile in children with parenteral nutrition-associated cholestasis[1], although it had previously been described in patients with hereditary phytosterolaemia (sitosterolaemia)[2]. More recently we observed a similar profile in 10 patients ascertained on the basis of the haematological phenotype of hereditary phytosterolaemia – a characteristic haemolytic anaemia with stomatocytes

Table 1 Techniques in regular use for selective screening of infants with liver disease

Detection of characteristic metabolites
Gas chromatography–mass spectrometry (GC-MS) of urine organic acids
e.g. to detect succinylacetone for the diagnosis of tyrosinaemia
Electrospray ionization (tandem) mass spectrometry ESI-MS(/MS) of urine bile acids
e.g. to detect 3-oxo-Δ^4 bile acids in Δ^4-3-oxosteroid 5β-reductase deficiency
ESI-MS/MS of blood spot carnitine species and amino acids
e.g. to detect C8-carnitine in medium-chain acyl-CoA dehydrogenase (MCAD) deficiency
e.g. to detect raised citrulline in citrullinaemia types I and II
Detection of abnormal proteins
Plasma protein electrophoresis with immunostaining
e.g. PiZ α_1-antitrypsin deficiency
e.g. transferrin glycoforms in congenital disorders of glycosylation
WBC enzymes
e.g. glucocerebrosidase for Gaucher disease
e.g. debranching enzyme for glycogen storage disease III
DNA
Blood
e.g. testing common mutations causing cystic fibrosis
Liver
e.g. testing for mitochondrial DNA depletion as a cause of liver failure

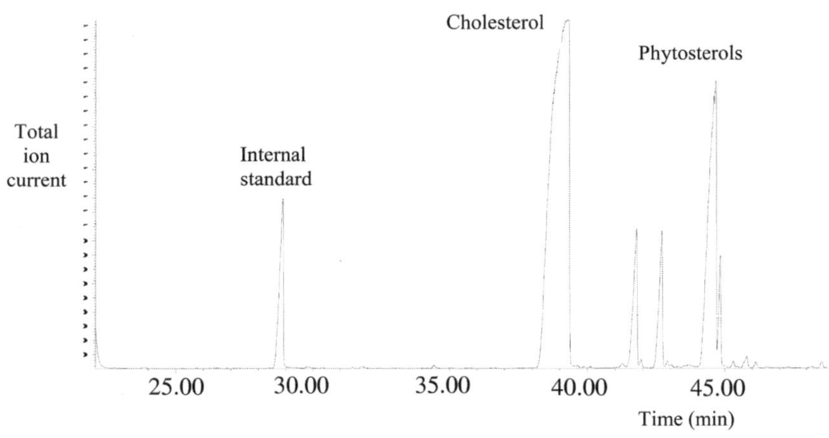

Figure 1 Analysis of plasma sterols by GC-MS: profile of a patient with hereditary phytosterolaemia (sitosterolaemia)

and macrothrombocytopenia[3]. These patients were shown to have mutations in the *ABCG5* and *ABCG8* genes – known to be the cause of hereditary phytosterolaemia[4]. Of the 10 individuals we described, two had raised transaminases and, when one of these individuals (an adult) was biopsied, the histology showed a severe non-alcoholic steatohepatosis (NASH) with grade 4 fibrosis – likely to require liver transplantion within 10 years. This led us to consider whether hereditary phytosterolaemia can cause liver disease. Hence we were very interested by the case report published by Miettinen et al. in February 2006[5]. They described a 19-year-old male with jaundice, splenomegaly, thrombocytopenia, leucopenia and mild haemolysis. Unconjugated bilirubin, conjugated bilirubin, transminases and alkaline phosphatase were all 2–3 times the upper limit of normal. IgG and IgM were raised and he had positive antinuclear and smooth muscle antibodies. A liver biopsy showed chronic active hepatitis and cirrhosis and a diagnosis of autoimmune chronic active hepatitis was made. However, 3 years later GC-MS of plasma sterols indicated that the patient had phytosterolaemia. Because of worsening liver function the patient required a liver transplant. Analysis of the *ABCG5* and *ABCG8* genes indicated he was a compound heterozygote for mutations in *ABCG8* (W361X and E423D). They pointed out that this was the first report of an association between hereditary sitosterolaemia and cirrhosis. They speculated that the cirrhosis might be due to a misfolded protein response or the effect of the accumulating phytosterols. Our experience also suggests that hereditary phytosterolaemia can cause cirrhosis.

GC-MS analysis of plasma sterols can also be useful in the diagnosis of inborn errors of cholesterol synthesis. Neonatal liver disease with jaundice and ascites can be a presenting feature of Smith–Lemli–Opitz syndrome (ref. 6 and unpublished observations) and lathosterolosis also causes cholestatic liver disease[7].

ANALYSIS OF URINE BILE ACIDS BY ELECTROSPRAY IONIZATION TANDEM MASS SPECTROMETRY

In 1986, when we introduced urine bile acid screening for infants with liver disease, the best technique available was fast atom bombardment mass spectrometry[8,9]. However, most laboratories now use electrospray ionization mass spectrometry[10,11]. This has the advantage of better detection of unconjugated bile acids (which tended to show unwanted fragmentation in the FAB source). This has, in our hands, led to improved detection of patients with amidation defects (unpublished observation). Table 2 lists the disorders of bile acid synthesis that cause neonatal cholestasis and can be detected by urine bile acid analysis[12].

GENETIC LIVER DISEASES DETECTABLE BY SCREENING OF CARNITINE SPECIES

The genetic liver diseases that can be diagnosed by analysis of electrospray ionization tandem mass spectrometry of blood spot acyl carnitine include: medium chain acyl-CoA dehydrogenase (MCAD) deficiency (↑ octanoyl carnitine [plus hexanoyl, C10:1]); very long chain acyl-CoA dehydrogenase (VLCAD) deficiency (↑ C14:1 carnitine); long chain 3-hydroxyacyl-CoA dehydrogenase (LCHAD) deficiency (↑ 3-OH C16 and C18:1 carnitine); carnitine palmitoyl transferase I deficiency (↑ free carnitine); and carnitine

Table 2 Disorders causing neonatal cholestasis and detectable by electrospray ionization tandem mass spectrometry of urine cholanoids (bile acids and bile alcohols)

Deficient enzyme	*Gene*	*Location*
3β-Hydroxy-Δ^5-C27 steroid dehydrogenase	*HSD3B7*	16p11.2-12
Δ^4-3-Oxosteroid-5β-reductase	*SRD5B1*	7q31
Sterol 27-hydroxylase (CTX)	*CYP27A1*	2q33-qter
Oxysterol 7α-hydroxylase	*CYP7B1*	8q21.3
α-Methyl-acyl-CoA racemase	*AMACR*	5p13.2-q11.1
Bile acid CoA: amino acid N-acyl transferase	*BAAT*	9q22.3
Peroxisome biogenesis	Several	

transporter deficiency (↓ free carnitine). MCAD deficiency is the commonest of these disorders and in the UK we are gaining experience of population screening for this disorder. Prior to screening our view of the disorder was that in the UK the incidence was 1 in 12 000 to 1 in 18 000 births, and that 85% of patients were homozygous for the A985G (Northern European) mutation. The disorder characteristically presented in the second year of life, at a median of 14 months. Following a common illness, children developed a Reye syndrome-like illness with vomiting, liver enlargement and encephalopathy leading in some cases to coma and in 20–25% of cases to death. Thirty per cent of the survivors had permanent brain damage. Following diagnosis patients were managed by ensuring continuing high carbohydrate intake (oral glucose polymer or intravenous 10% dextrose) during intercurrent illness. Among 41 patients followed up for a median of 6.7 years, only two had episodes of encephalopathy and there were no deaths or chronic morbidity[13]. Thus, for MCAD deficiency, there was a definite latent period (apart from rare neonatal presentations), a blood spot marker – C8 carnitine concentration, a known natural history and response to treatment. MCAD was therefore considered a good candidate for neonatal screening.

The results of the UK MCAD screening project will be published elsewhere, but from the point of view of a metabolic paediatrician in north London it is clear that we are detecting more cases than we expected and that most of these 'extra' cases are not of Northern European descent and do not have the A985G mutation. The natural history of some of these MCAD variants is not known. Similar results were reported by Wilcken et al. from a screening programme in New South Wales and the Australian Capital Territory, Australia[14]. We are reassured by the fact that, in line with previous follow-up studies of diagnosed MCAD patients, there have been no deaths and there has been no significant morbidity in screen-positive infants treated in our centre.

Some of the 'disorders' that can be detected by detailed analysis of the blood spot acylcarnitine profile should not be regarded as diseases at all – they so rarely give rise to symptoms. This is the case with 3-methylcrotonylglycinuria (3-methylcrotonyl-CoA carboxylase deficiency, MCCD). The low ratio of benefit to harm was pivotal to the decision to exclude MCCD from the newborn screening programme in Germany[15].

POTENTIAL OF HPLC–ESI–MS/MS OF METABOLITES

It has become clear that ESI–MS/MS of metabolites, sometimes coupled to high-performance liquid chromatography (HPLC) is a very versatile method for the detection of a group of important metabolites. This can facilitate the diagnosis of known disorders and also allow the detection of new disorders.

In our laboratories the detection of increased hydroxy-C4 carnitine in blood spots led to the description of the first case of short chain 3-hydroxyacyl-CoA dehydrogenase deficiency[16] and the second case of hydroxyisobutyryl-CoA hydrolase deficiency[17]. HPLC–ESI–MS/MS could be used for analysis of α-aminoadipic semialdehyde when we showed that this was an important metabolite for the diagnosis of pyridoxine-dependent epilepsy[18].

Not only does ESI–MS/MS allow the detection of many metabolites in one short assay, but it can also be used to perform multiplex enzyme assays[19–21]. Li, Wang and co-workers have devised methods for the measurement of the activities of several lysosomal enzymes in dried blood spots. Substrates and products are detected by tandem mass spectrometry and the assays are conducted simultaneously (multiplexing). Complete separation from controls and heterozygotes has been achieved for the following disorders: mucopolysaccharidosis I (Hurler disease), Gaucher disease, Niemann–Pick disease types A and B, Fabry disease, Krabbe disease, Pompe disease and mucopolysaccharidosis type II (Hunter disease). Enzyme replacement therapy is available for Gaucher disease, Fabry disease, Pompe disease and Hurler disease (as a bridge to bone marrow transplantation) and will soon be available for Hunter disease. However, treatment is expensive, there is uncertainty (for some disorders) as to which patients need treatment. We have little information in most cases of the effect on the phenotype of treatment from the neonatal period.

PROTEIN/PEPTIDE MASS SPECTROMETRY

If the advances in the technology available for metabolite screening are impressive, the advances in protein and peptide analysis by mass spectrometry are truly stunning. Are any of these advances likely to have a role in clinical diagnosis or population screening screening?

Methods using 2D-PAGE

Most of the early methods used in proteomics involved separation of proteins by two-dimensional polyacrylamide gel eletrophoresis (2D-PAGE) and analysis of peptides released by in-gel digestion of protein spots by matrix-assisted laser desorption ionization time-of-flight mass spectrometry (MALDI-TOF MS). We have shown that, using these techniques on plasma samples, it is possible to detect α_1-antitrypsin variants be they due to altered glycosylation or changes in amino acid sequence (e.g. the E342K sequence change in the PiZ variant)[22]. Thus, there is no doubt that it is possible to diagnose genetic liver diseases at the protein level using these techniques.

The scope of proteomic techniques has been extended by methods that allow direct quantitative comparison between a sample and controls e.g. two-dimensional difference gel electrophoresis (2D-DIGE), in which the proteins from sample and controls are tagged with different fluorescent markers. Another major advance has been the introduction of quadrupole-time-of-flight (Q-TOF) mass spectrometers that can directly read the amino acid sequence of a peptide[23].

In theory, many patients with genetic liver diseases could be diagnosed by proteomic methodology provided that: (a) the affected protein is expressed in the matrix being examined, e.g. plasma, white cells, liver biopsy; (b) the methods in use can determine the amount of protein present; (c) peptide mass mapping gives complete coverage of the protein sequence; and (d) any abnormal peptide is present in sufficient amount for sequencing by Q-TOF analysis. (It is important to remember that some amino acid changes will alter the electrophoretic mobility of the protein; others will only affect the peptide mass map.) However, because of the skill required to get good results from 2D-PAGE and the large amount of time that is required to comprehensively analyse a single sample, it is likely that these will remain research techniques. For screening, we need to develop different strategies for the detection of abnormal proteins, strategies that involve minimal sample work-up and rapid analysis.

Methods with simplified sample work-up

One approach that has been used successfully in haemoglobinopathy screening has been the generation of a tryptic digest of all the proteins eluted from a blood spot and then anlaysing this using a triple quadrupole mass spectrometer in multiple reaction monitoring (MRM) mode[24]. In this way specific abnormal peptides can be sought on the basis of their mass and fragmentation pattern. It remains to be seen whether this sort of method will work on proteins of interest in the diagnosis of genetic liver diseases (which are present in much smaller amounts than haemoglobin in a blood spot). The method requires the operator to know the specific protein and amino acid change he is looking for, and this is another major limitation.

A second approach has been to use surface chemistry technology to selectively bind certain proteins in a matrix and then to analyse those proteins that have been adsorbed (as intact proteins) using laser desorption time-of-flight mass spectrometry. This is called surface-enhanced laser desorption ionization (SELDI) TOF-MS. A selection of different surfaces can be used for a single sample producing protein arrays. The simplest surfaces are a cation ion exchange surface for positively charged proteins, an anion ion exchange surface for negative proteins, immobilized metal ions for metal-binding proteins and phosphoproteins and a reverse phase surface for hydrophobic proteins such as membrane-bound proteins. However, it is also possible to bind on to a surface: an antibody to selectively adsorb the target protein, a physiological/pharmacological ligand for a protein, e.g. a drug to capture a receptor protein, or a substrate to capture an enzyme or transporter. Proteins can be bound in order to use known protein–protein interactions and DNA can be bound to a surface to look for a transcription factor which will bind to that DNA sequence. The power of the SELDI-TOF MS technique is that the clean-up on the chip allows for the detection of minor as well as major proteins. The software of the instrument allows rapid comparison of protein profiles in a specific disorder with controls. This is an attractive way of trying to detect biomarkers for a particular disorder. Having found a marker, the chip technology can be scaled up to isolate and identify protein of interest.

We have shown that SELDI-TOF MS analysis with anti-transferrin antibody on the chip can be used to diagnose inborn errors of protein glycosylation (several of which produce liver disease)[23].

CRITERIA FOR POPULATION SCREENING

An abbreviated version of the criteria for neonatal screening drawn up by the UK National Neonatal Screening Committee is shown in Table 3. It is important to recognize that:

Table 3 Criteria that should be met before introduction of neonatal screening for a disorder – as proposed by the UK National Neonatal Screening Committee (abbreviated version)

The condition
Important health problem
Epidemiology and natural history adequately understood
Disease marker and latent period
Cost-effective primary preventions implemented
The test
Simple, safe, precise, validated
Suitable cut-off level defined and agreed
Test acceptable to the population
Agreed policy on further diagnostic investigation
The treatment
Effective: evidence that early treatment leads to better outcomes than late treatment
Agreed policies on who to treat and how to treat
Clinical management and patient outcomes should be optimized before screening
The screening programme
Evidence that programme reduces mortality and morbidity
Evidence that programme is clinically, socially, ethically acceptable to health professionals and public
Programme should outweigh physical and psychological harm of the test
Cost economically balanced

1. The experience with screening for MCAD deficiency and MCCD deficiency indicates that screening programmes may detect individuals with mild disease or no disease, in other words a pilot screening programme may be an important part of determining the true range of the natural history of the disorder when it is ascertained by a screening programme.

2. These criteria are not based purely on the best possible medical practice but are drawn up in the context of limited resources. Hence phrases such as 'important health problem' and 'cost economically balanced'. Screening for bile acid synthesis defects could easily be justified in terms of the improved outcome for the affected infants but, because they are rare, it would be possible to argue that they are not an 'important health problem'. On the other hand, treatment is very cheap compared to the treatment of a lysosomal disorder by enzyme replacement therapy. In the area of neonatal screening difficult political choices have to be made.

CONCLUSIONS

As a result of advances in mass spectrometry there will be a steady increase in the number of metabolites that can be measured in blood spots. A number of these will find application in neonatal screening for treatable genetic liver diseases.

Many proteins will be quantifiable; for some specific changes in amino acid sequence will be detectable. Some of the emerging techniques for analysis of proteins in blood spots have potential for screening for genetic liver diseases.

Multicentre collaboration on pilot projects will be required to prove the benefits of screening for rare disorders.

References

1. Clayton PT, Bowron A, Mills KA, Massoud A, Casteels M, Milla PJ. Phytosterolaemia in children with parenteral nutrition-associated cholestatic liver disease. Gastroenterology. 1993;105:1806–13.
2. Bhattacharyya AK, Connor WE. Beta-sitosterolemia and xanthomatosis. A newly described lipid storage disease in two sisters. J Clin Invest. 1974;53:1033–43.
3. Rees DC, Iolascon A, Carella M et al. Mediterranean stomatocytosis/macro-thrombocytopenia and phytosterolemia. Br J Haematol. 2005;130:297–309.
4. Berge KE, Tian H, Graf GA et al. Accumulation of dietary cholesterol in sitosterolemia caused by mutations in adjacent ABC transporters. Science. 2000;290:1771–5.
5. Miettinen TA, Klett EL, Gylling H, Isoniemi H, Patel SB. Liver transplantation in a patient with sitosterolemia and cirrhosis. Gastroenterology. 2006;130:542–7.
6. Rossi M, Vajro P, Iorio R et al. Characterization of liver involvement in defects of cholesterol biosynthesis: long-term follow-up and review. Am J Med Genet A. 2005;132: 144–51 (Review).
7. Brunetti-Pierri N, Corso G, Rossi M et al. Lathosterolosis, a novel multiple-malformation/mental retardation syndrome due to deficiency of 3beta-hydroxysteroid-delta5-desaturase. Am J Hum Genet. 2002;71:952–8. (Erratum in: Am J Hum Genet. 2003;73:445).
8. Lawson AM, Madigan MJ, Shortland DB, Clayton PT. Rapid diagnosis of Zellweger syndrome and infantile Refsum disease by fast atom bombardment of urine bile salts. Clin Chim Acta. 1986;161:221–31.

9. Clayton PT, Leonard JV, Lawson AM et al. Familial giant cell hepatitis associated with synthesis of 3β,7α-dihydroxy- and 3β,7α,12α-trihydroxy-5-cholenoic acids. J Clin Invest. 1987;79:1031–8.
10. Mills KA, Mushtaq I, Johnson AW, Whitfield PD, Clayton PT. A method for the quantitation of conjugated bile acids in dried blood spots using electrospray ionization-mass spectrometry. Pediatr Res. 1998;43:361–8.
11. Yousef IM, Perwaiz S, Lamireau T, Tuchweber B. Urinary bile acid profile in children with inborn errors of bile acid metabolism and chronic cholestasis; screening technique using electrospray tandem mass-spectrometry (ES/MS/MS). Med Sci Monit. 2003;9:MT21–31.
12. Clayton PT. Disorders of bile acid synthesis. In: Fernandes J, Saudubray J-M, Van den Berghe G, Walter JH, editors. Inherited Metabolic Diseases. Diagnosis and Treatment, 4th edn. Berlin: Springer-Verlag, 2006:421–30.
13. Wilson CJ, Champion MP, Collins JE, Clayton PT, Leonard JV. Outcome of medium chain acyl-CoA dehydrogenase deficiency after diagnosis. Arch Dis Child. 1999;80:459–62.
14. Wilcken B, Wiley V, Hammond J, Carpenter K. Screening newborns for inborn errors of metabolism by tandem mass spectrometry. N Engl J Med. 2003;348:2304–12.
15. Stadler SC, Polanetz R, Maier EM et al. Newborn screening for 3-methylcrotonyl-CoA carboxylase deficiency: population heterogeneity of MCCA and MCCB mutations and impact on risk assessment. Hum Mutat. 2006;27:748–59.
16. Clayton PT, Eaton S, Aynsley-Green A et al. Hyperinsulinism in short-chain L-3-hydroxyacyl-CoA dehydrogenase deficiency reveals the importance of beta-oxidation in insulin secretion. J Clin Invest. 2001;108:457–65.
17. Loupatty FJ, Clayton PT, Ruiter JP et al. Mutations in the gene encoding 3-hydroxyisobutyryl-CoA hydrolase results in progressive infantile neurodegeneration. Am J Hum Genet. 2007;80:195–9.
18. Mills PB, Struys E, Jakobs C et al. Mutations in antiquitin in individuals with pyridoxine-dependent seizures. Nat Med. 2006;12:307–9.
19. Li Y, Scott CR, Chamoles NA et al. Direct multiplex assay of lysosomal enzymes in dried blood spots for newborn screening. Clin Chem. 2004;50:1785–96.
20. Wang D, Eadala B, Sadilek M et al. Tandem mass spectrometric analysis of dried blood spots for screening of mucopolysaccharidosis I in newborns. Clin Chem. 2005;51:898–900.
21. Wang D, Wood T, Sadilek M, Scott CR, Turecek F, Gelb MH. Tandem mass spectrometry for the direct assay of enzymes in dried blood spots: application to newborn screening for mucopolysaccharidosis II (Hunter disease). Clin Chem. 2007;53:137–40.
22. Mills K, Mills PB, Clayton PT, Johnson AW, Whitehouse DB, Winchester BG. Identification of alpha(1)-antitrypsin variants in plasma with the use of proteomic technology. Clin Chem. 2001;47:2012–22.
23. Mills K, Mills P, Jackson M et al. Diagnosis of congenital disorders of glycosylation type-I using protein chip technology. Proteomics. 2006;6:2295–304.
24. Daniel YA, Turner C, Haynes RM, Hunt BJ, Dalton RN. Rapid and specific detection of clinically significant haemoglobinopathies using electrospray mass spectrometry–mass spectrometry. Br J Haematol. 2005;130:635–43.

15
Gene therapy for liver diseases: experimental strategies

H. E. BLUM

INTRODUCTION

Molecular biology and recombinant DNA technology increasingly contribute to the diagnosis, therapy and prevention of human diseases. Molecular methods allow the early and/or specific detection of inherited, infectious and malignant liver diseases. In addition, such analyses increasingly led to a better understanding of the pathogenesis of the various liver diseases, which in turn had an impact on patient management, including the presymptomatic identification of patients at risk, the correct staging of the disease and the follow-up of patients undergoing therapy. Thus, molecular biology is increasingly becoming an integral part of basic as well as clinical hepatology. This chapter will briefly review current concepts and potential applications of gene therapy for the treatment or prevention of various liver diseases.

Genetic classification of diseases

Genetically, human diseases can be classified into three major categories[1–4]:

1. Hereditary monogenic diseases that are caused by a single gene defect and inherited by the classical Mendelian rules. There are more than 4000 monogenic diseases described. For an increasing number of these diseases the genetic basis is being identified.

2. Acquired monogenic diseases that are infections as well as malignancies which are caused by the mutation or epigenetic modification of a single gene.

3. Complex genetic diseases that are associated with mutations of several genes which are acquired and frequently accumulated during a lifetime. Several common human diseases belong to this category, such as most malignancies.

Gene therapy is defined as the introduction of genetic material into human cells with a therapeutic or preventive benefit. In a broader definition, cell or organ transplantation are included. This chapter will discuss the basic concepts of gene therapy[1–7] as well as some therapeutic and preventive applications for liver diseases.

GENE THERAPY OF LIVER DISEASES

Based on the genetic classification of diseases detailed above, the principle of gene therapy involves six therapeutic concepts (Table 1): gene repair, gene substitution or cell therapy for hereditary monogenic diseases, block of gene expression or DNA vaccination for acquired monogenic diseases, and gene augmentation or DNA vaccination for complex genetic diseases. For clinical applications, gene therapy is explored with the aim of either providing novel therapeutic strategies for diseases for which there is no treatment available, or replacing and in some cases complementing existing treatment modalities, thereby increasing therapeutic efficacy and/or reducing adverse events.

Table 1 Concepts of gene therapy

Diseases	*Gene therapy*
Hereditary monogenic diseases	Gene repair Gene replacement Cell or organ transplantation
Acquired monogenic diseases	Block of gene expression DNA vaccination
Complex genetic diseases	Gene augmentation DNA vaccination

Gene repair

An increasing number of liver diseases has been molecularly defined as a defect of a single gene (Table 2). In this context, one therapeutic concept is the *in vitro* or *in vivo* repair of the defective gene. In the Gunn rat model of Crigler–Najjar syndrome type I Kren et al.[8] were able to partially correct the genetic defect underlying the UDP-glucuronosyl transferase deficiency by the intravenous injection of a cyclic normal/wild-type chimeric oligonucleotide. While these findings have not been independently confirmed or extended to other hereditary monogenic (liver) diseases, the data suggest that it is in principle possible to repair a gene defect *in vivo*. Further, it has been shown that cellular RNA species can be modified by *trans*-splicing group I ribozymes. Such ribozymes may in principle allow one to treat a variety of inherited diseases at the RNA level[9–11].

Table 2 Hereditary monogenic liver diseases (selection)

Gene	*Disease*
UDP-glucuronosyl transferase	Crigler–Najjar syndrome type I
Alpha-1-antitrypsin	Liver cirrhosis, emphysema
CF transmembrane regulator	Mucoviscidosis, cystic fibrosis
Factor VIII	Haemophilia A
Factor IX	Haemophilia B
Fumarylacetoacetate hydrolase	Tyrosinaemia type 1
LDL receptor	Familial hypercholesterolaemia
Ornithine transcarbamylase	Hyperammonaemia

Gene substitution

The targeted substitution of a defective cellular gene by the normal/wild-type homologue with production of the physiological gene product is another approach to correct a hereditary or acquired monogenic gene defect. In an animal model of hereditary tyrosinaemia type 1 (HT1), a liver disease caused by a deficiency of fumarylacetoacetate hydrolase (FAH), multiple injections of a retroviral vector carrying the FAH gene resulted in a gene transfer efficiency of >90% to hepatocytes and the restoration of a normal liver function[12,13]. In patients, examples for gene substitution are the partial correction of severe haemophilia A by the *ex vivo* transduction of autologous skin fibroblasts with the normal/wild-type factor VIII gene, followed by laparoscopic implantation of the genetically modified fibroblasts into the omentum majus[14], or of haemophilia B by adenovirus-associated vector (AAV)-based gene transfer[15].

In rare situations in which a hepatocelluar carcinoma (HCC) is caused by the mutation of a tumour-suppressor gene, e.g. the p53 gene, the substitution of the mutated by the normal/wild-type gene *in vitro* can reduce the number of tumour cell colonies and restore cisplatin sensitivity[16,17].

Cell therapy

Allogeneic or *ex vivo* genetically modified autologous hepatocyte transplantation is a promising strategy to treat hereditary monogenic liver diseases. In patients with familial hypercholesterolaemia (FH) that is caused by various mutations in the low-density lipoprotein (LDL) receptor gene[18], apart from orthotopic liver transplantation[19,20], liver-directed gene therapy has been performed in a pilot study in five patients[21,22]. Autologous liver cells, prepared from a surgical biopsy, were transduced *ex vivo* with a recombinant retrovirus expressing the normal LDL receptor. These *ex vivo* genetically modified hepatocytes were transplanted by portal infusion and resulted in significant and prolonged reductions in LDL cholesterol in three of five patients for at least 4 months, demonstrating the feasibility of engrafting a limited number of *ex vivo* transduced hepatocytes. Allogeneic hepatocyte transplantation has also been successfully used in patients to partially correct Crigler–Najjar syndrome type I[23] and glycogen storage disease type I[24].

Block of gene expression or function

For diseases caused by the expression of an acquired gene or the overexpression of an endogenous gene, blocking gene expression can be an effective therapeutic approach. Several strategies can be employed (Figure 1): interference with the transcription of genes by binding of transcription factors to nucleic acids introduced into or synthesized in the cells (decoy strategy)[25,26]; by binding of single-stranded nucleic acids to double-stranded DNA, forming a triple helix structure[25,26]; hybridization of RNA molecules possessing endonuclease activity (ribozymes) to RNA, resulting in its sequence-specific cleavage[27,28]; RNA interference (RNAi) by small inhibiting RNA (siRNA) or microRNA (miRNA)[29–32]; block of translation by antisense oligonucleotides[25,26,33,34] and the intracellular synthesis of peptides or proteins, interfering with their normal counterpart, termed a dominant negative (DN) mutant strategy[35]. These different strategies have been applied to a number of malignant and infectious diseases. In particular ribozymes, siRNA, antisense oligonucleotides and DN mutants have been experimentally explored to treat hepatitis B virus (HBV) and hepatitis C virus (HCV) infections.

Ribozymes

Ribozymes ('ribonucleic acid enzymes') were originally discovered as naturally occurring RNA molecules that catalyse the sequence-specific cleavage of RNA and RNA splicing reactions[27,28]. This catalytic activity is the major attraction of the ribozyme concept since one ribozyme can cleave many target RNA molecules. Ribozymes that cleave RNA are being developed as inhibitors of gene expression and viral replication. Several studies have clearly demonstrated that hammerhead ribozymes can specifically cleave HBV RNA[36,37] or HCV RNA[38,39] *in vitro*. *In vivo*, however, an efficient ribozyme-mediated cleavage of HBV RNA could not be demonstrated to date. For HCV infection, the elimination of HCV RNA in infected hepatocytes by ribozymes has also been reported[38,40].

Small interfering RNA

RNAi is a recently discovered basic intracellular mechanism[29–32] that has also been explored for the inhibition of HBV and HCV infection (Figure 2). For HBV, inhibition of viral gene expression and replication has been shown *in vitro*[41–43] and in different mouse models *in vivo*[44–48]. For HCV, inhibition of viral gene expression and replication has been shown *in vitro* in the replicon system[49–51]. While effective in blocking viral gene expression and replication, *in vivo* oversaturation of celluar miRNA/short hairpin RNA (shRNA) pathways can result in lethal hepatotoxicity[48]. For future RNAi-based strategies in animals or humans, these findings indicate that the control of intracellular shRNA expression levels through optimizing shRNA dose and sequence will be key to reduce the risk of oversaturating endogenous small RNA pathways.

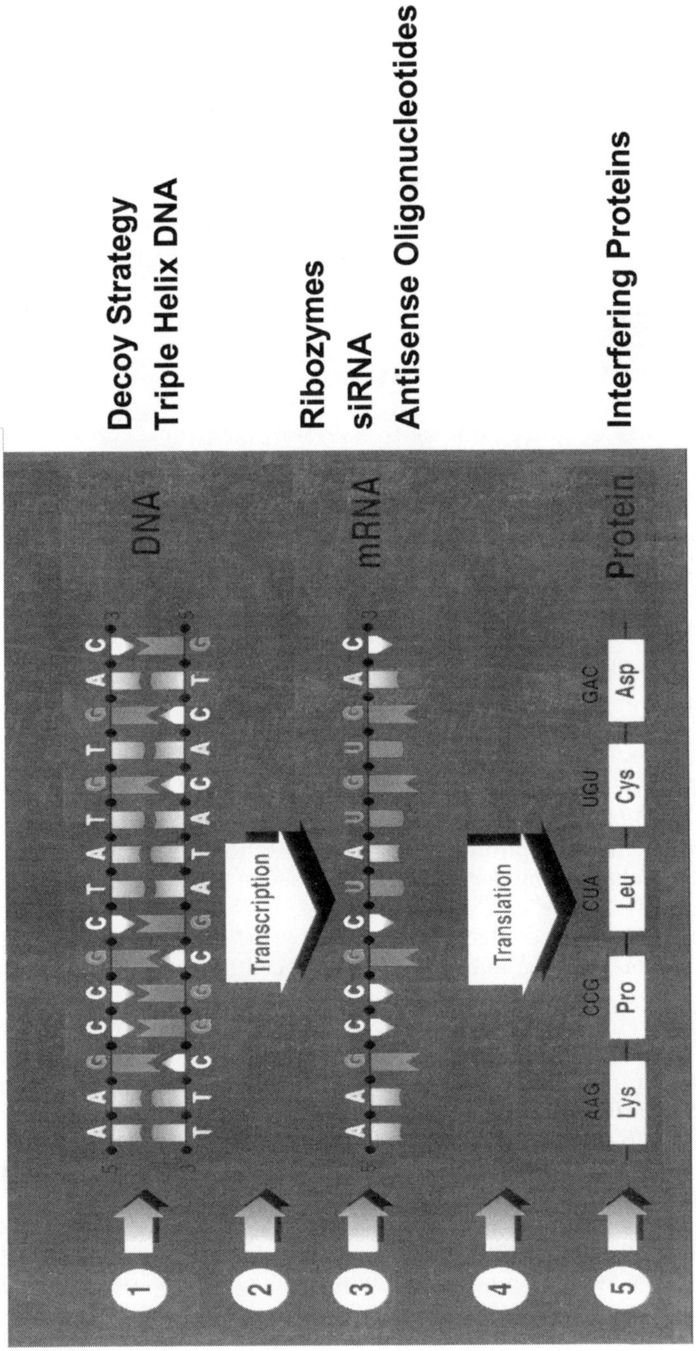

Figure 1 Strategies aimed at blocking gene expression

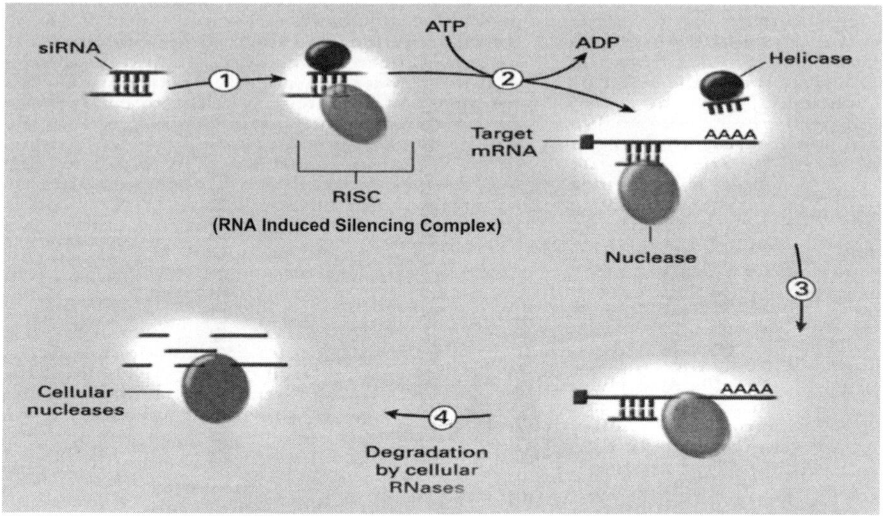

Figure 2 Principle of RNA interference[31]

Antisense oligonucleotides

Antisense nucleic acids are designed to specifically bind to RNA or mRNA, resulting in the formation of RNA–DNA (antisense oligodeoxynucleotides) or RNA–RNA hybrids (antisense oligoribonucleotides) with an arrest of RNA replication, reverse transcription or mRNA translation[25,26,33,34,52]. Antisense effects can be potentiated by degradation of RNA in RNA–DNA hybrids by cellular RNases H. While conceptually simple, it is clear now that not all desired as well as undesired effects are caused by the target sequence-specific antisense action of the oligonucleotides or the cellular enzymes mentioned above[53,54].

The antisense strategy has been successfully applied *in vitro* to HBV infection[55–58] and to HCV infection[59–64]. In addition, studies in nude mice[65], in the duck hepatitis B virus (DHBV)[66] and the woodchuck hepatitis virus (WHV) model of HBV infection[67] demonstrated the *in vivo* applicability of this approach. While no toxic effects have been observed in these experiments, the contribution of non-antisense effects to the inhibition of viral replication or gene expression has not been systematically assessed in most studies. Independent of the antisense or non-antisense mechanism of the biological effects, an *in vitro* screening procedure for the identification of functionally active oligonucleotides[53,68] should greatly facilitate the design of oligonucleotide-based antiviral therapies.

Interfering peptides or proteins

The intracellular synthesis of interfering peptides or proteins, including single-chain or whole non-secreted antibodies, is aimed at the specific interference with the assembly or function of viral structural or non-structural proteins and represents a type of intracellular immunization[69]. This approach has been shown for block mammalian and avian hepadnavirus gene expression and replication *in vitro*. For example, the fusion of different polypeptides of various lengths to the carboxy-terminus of the viral core protein yields DN mutants[70–73]. These DN mutants are species-specific and suppress viral replication by at least 90% at an effector to target ratio of 1:10. Moreover, the non-secretory form of the hepatitis B e antigen (HBeAg) was shown to effectively inhibit viral replication and may indeed act as a natural regulator of HBV propagation[74–76]. The potential advantage of DN mutants over ribozymes or antisense oligonucleotides is their relative independence from viral sequence variations, minimizing the risk of selecting or accumulating 'therapy escape' mutants.

DNA vaccination

A novel approach is DNA vaccination resulting in the manipulation of the immune system by introduction of expression vectors into muscle cells or dendritic cells and long-lasting cellular and humoral immune responses. Direct gene transfer into muscle[77] represents an exciting new development and elegant application of gene therapy[78,79]. The therapeutic DNA vaccine acts by the intracellular plasmid-derived synthesis of a viral protein which enters the cell's MHC class I pathway[78]. The proteins originating within the cell can be processed by MHC class I molecules that carry fragments of the protein to the cell surface. There they stimulate $CD8^+$ cytotoxic T cells, resulting in cell-mediated immunity. In principle this strategy is applicable to the treatment of acquired genetic diseases, associated with the expression of disease-specific antigens serving as targets for $CD8^+$ cytotoxic T cells.

Therapeutic DNA vaccination has been experimentally explored for HBV[80–84] as well as HCV infection[85,86], and holds great promise as an effective molecular therapy for these viral diseases. In this context the coexpression of HBsAg and interleukin-2 was shown to greatly increase humoral as well as cellular immune response[87].

Further, DNA-based tumour vaccination against HCC may be possible, for example, by intramuscular introduction of a plasmid expressing HCC-specific antigens or antigens that are highly overexpressed in HCC cells, such as AF-20 antigen, insulin receptor substrate-1[88], alpha-fetoprotein[89], aspartyl asparaginyl hydroxylase, mutated p53 protein and others. Potential limitations of this strategy include the regulation of the immune response as well as the low-level expression of the targeted antigen in non-malignant cells[90], rendering them susceptible to immune-mediated elimination as well.

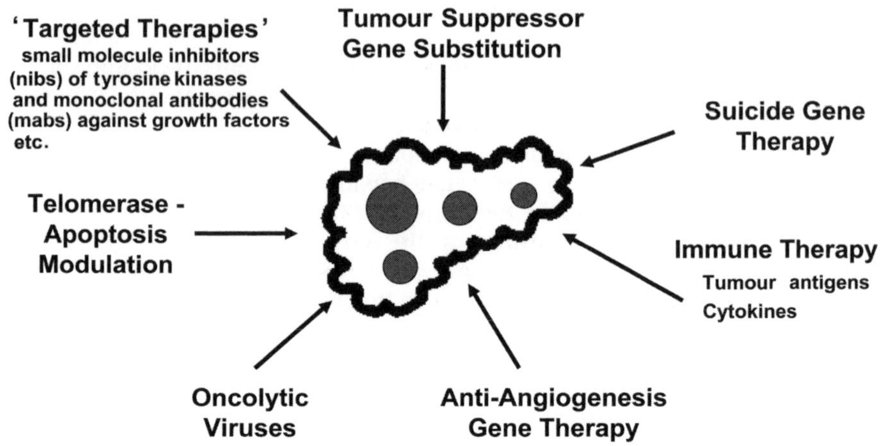

Figure 3 HCC treatment: experimental strategies, including gene therapy

Gene augmentation

Complex genetic diseases are among the most prevalent clinical problems. In this situation, gene augmentation is aimed at the local expression of a therapeutic gene product that is physiologically not expressed or expressed at therapeutically insufficient levels. This strategy is explored among others for the treatment of HCC (Figure 3).

Suicide gene therapy

An interesting strategy to treat HCC is genetic prodrug activation therapy via the introduction of a 'suicide gene' into malignant cells followed by the administration of the prodrug. This concept has been experimentally explored in HCC cells *in vitro* and *in vivo*, e.g. for the HSV-tk gene[91–95], the gene encoding cytosine deaminase (CD) that converts the prodrug 5-fluorocytosine to 5-fluorouracil which inhibits RNA and DNA synthesis during the S-phase of the cell cycle[96], the gene encoding purine nucleoside phosphorylase that converts purine analogues into freely diffusible toxic metabolites[97,98], as well as the gene encoding cytochrome P450 4B1[99]. A significant bystander effect of cell killing caused by suicide gene expression could be demonstrated *in vitro* and *in vivo*, based on cell–cell contact rather than release of cytotoxic substances from the transduced cells[100]. At the same time the bystander effect may also affect non-malignant dividing cells in the target tissue, potentially limiting the application of this strategy.

Immune therapy

In the process of malignant transformation new antigenic surface proteins can be expressed (tumour antigens) or oncofetal proteins can be re-expressed, e.g., alpha-fetoprotein (AFP).

AFP-specific immune therapy has been explored in mice and humans. Vaccination with an AFP-expressing DNA construct resulted in tumour rejection and prolonged survival in a mouse model[89]. In patients AFP-specific T cells could also be detected[101,102]. Since AFP is expressed not only by tumour cells but also by regenerating liver cells and in liver cirrhosis immunization against AFP carries the risk of autoimmune hepatitis which has been experimentally shown in mice[90].

Immune therapy with antigen-presenting cells (APC) is another strategy that has been explored using dendritic cells (DC) exposed to tumour lysates, peptides or *ex vivo* tranduced with tumour antigen-expressing DNA constructs. While this strategy is conceptually very interesting, to date there are no data available that demonstrate its clinical efficacy[103].

Cytokine gene therapy has been explored using tumour necrosis factor (TNF)-alpha, GM-CSF, interferon-alpha or -gamma, interleukin (IL)-2, -4, -6, -7, -12 and -18, B7-1 as well as CD40 ligand. Complete regression of a HCC was demonstrated *in vivo* by TNF-alpha[104], IL-2[105], IL-12[106] and an activatable interferon regulatory factor-1 in mice[107]. Gene transfer was achieved *in vivo* by delivering retroviral[104] or adenoviral vectors[105] systemically, directly into the tumour or into the peritoneal cavity. A pilot study in patients with gastrointestinal tumours exploring the intratumoral injection of an adenoviral IL-12 expression construct showed only marginal efficacy, however[108].

Antiangiogenic gene therapy

This concept has been experimentally explored in a HCC mouse model using the angiostatin gene. Angiostatin gene transfer resulted in reduced tumour volume and vascular density[109].

Oncolytic viruses

This new and elegant approach uses p53 mutations for selective, adenovirus-mediated lysis of tumour cells. Thus, an adenovirus mutant was engineered that replicates selectively in p53-deficient human tumour cells[110–112]. Other examples are the adenoviral introduction of Smac which antagonizes the inhibitor of apoptosis proteins in HCC tumour cells and enhances tumour cell death[113] as well as tumour-specific replication-restricted adenoviral vectors[114]. Further, the intravascular administration of a replication-competent genetically engineered herpes simplex virus (HSV)-1 resulted in oncolysis of a diffuse HCC[115]. More efficient HSV-1-based vectors have been developed[116].

MOLECULAR PREVENTION OF LIVER DISEASES

DNA-based prophylactic vaccination against HBV infection, for example, is possible by intramuscular introduction of a plasmid expressing hepatitis B surface antigen (HBsAg). HBsAg is taken up by cells via phagocytosis or endocytosis, processed through the major histocompatibility complex (MHC) class II system and primarily stimulates an antibody response through $CD4^+$ helper T cells with the production of anti-HBs[78,79,117–120]. While the DNA-based vaccination against HBV infection induces anti-HBs antibodies and prevents HBV infection, DNA-based vaccination against HCV infection of chimpanzees has been shown not to prevent infection but to result in the resolution of acute HCV infection through an effective vaccine-induced cellular immune response[121,122].

SUMMARY AND PERSPECTIVES

Molecular analyses have become an integral part of biomedical research as well as clinical medicine. The definition of the genetic basis of many human diseases has led to a better understanding of their pathogenesis, and has in addition offered new perspectives for their diagnosis, therapy and prevention. Genetically, human diseases can be classified as hereditary monogenic, acquired monogenic and complex genetic diseases. Based on this classification, gene therapy is based on six concepts: gene repair, gene substitution, cell or organ transplanation, block of gene expression or function, gene augmentation and DNA vaccination. While the recent developments in gene therapy for liver diseases are promising, various delivery, targeting and safety issues need to be addressed before these strategies can enter clinical practice. Nevertheless, gene therapy will become part of the management of patients with liver diseases, complementing existing diagnostic, therapeutic and preventive strategies.

References

1. Anderson WF. Human gene therapy. Science. 1992;256:808–13.
2. Friedmann T. Progress toward human gene therapy. Science. 1989;244:1275–81.
3. Gutierrez AA, Lemoine NR, Sikora K. Gene therapy for cancer. Lancet. 1992;339:715–21.
4. Miller AD. Human gene therapy comes of age. Nature. 1992;357:455–60.
5. Morgan RA, Anderson WF. Human gene therapy. Annu Rev Biochem. 1993;62:191–217.
6. Mulligan RC. The basic science of gene therapy. Science. 1993;260:926–32.
7. Rosenberg SA. Gene therapy for cancer. J Am Med Assoc. 1992;268:2416–19.
8. Kren BT, Parashar B, Bandyopadhyay P, Chowdhury NR, Chowdhury JR, Steer CJ. Correction of the UDP-glucuronosyltransferase gene defect in the Gunn rat model of Crigler–Najjar syndrome type I with a chimeric oligonucleotide. Proc Natl Acad Sci USA. 1999;96:10349–54.
9. Sarver N, Cairns S. Ribozyme *trans*-splicing and RNA tagging: following the messenger. Nat Med. 1996;2:641–42.
10. Jones JT, Lee SW, Sullenger BA. Tagging ribozyme reaction sites to follow *trans*-splicing in mammalian cells. Nat Med. 1996;2:643–8.
11. Lan N, Howrey RP, Lee S-W, Smith CA, Sullenger BA. Ribozyme-mediated repair of sickle β-globin mRNAs in erythrocyte precursors. Science. 1998;280:1593–6.

12. Overturf K, Al-Dhalimy M, Tanguay R et al. Hepatocytes corrected by gene therapy are selected *in vivo* in a murine model of hereditary tyrosinaemia type I. Nat Genet. 1996;12: 266–73.
13. Grompe M. The pathophysiology and treatment of hereditary tyrosinemia type 1. Semin Liver Dis. 2001;21:563–71.
14. Roth DA, Tawa NE Jr, O'Brien JM, Treco DA, Selden RF. Nonviral transfer of the gene encoding coagulation factor VIII in patients with severe hemophilia A. N Engl J Med. 2001;344:1735–42.
15. Kay MA, Manno CS, Ragni MV et al. Evidence for gene transfer and expression of factor IX in haemophilia B patients treated with an AAV vector. Nat Genet. 2000;24:257–61.
16. Xu GW, Sun ZT, Forrester K, Wang XW, Coursen J, Harris CC. Tissue-specific growth suppression and chemosensitivity promotion in human hepatocellular carcinoma cells by retroviral-mediated transfer of the wild-type p53 gene. Hepatology. 1996;24:1264–8.
17. Zender L, Kock R, Eckhard M et al. Gene therapy by intrahepatic and intratumoral trafficking of p53-VP22 induces regression of liver tumors. Gastroenterology. 2002;123: 608–18.
18. Brown MS, Goldstein JL. A receptor-mediated pathway for cholesterol homeostasis. Science. 1986;232:34–7.
19. Bilheimer DW, Goldstein JL, Grundy SM, Starzl TE. Liver transplantation to provide low density lipoprotein receptors and lower plasma cholesterol in a child with homozygous familial hypercholesterolemia. N Engl J Med. 1984;311:1658–64.
20. Hoeg JM, Starzl TE, Brewer HB. Liver transplantation for treatment of cardiovascular disease: comparison with medication and plasma exchange in homozygous familial hypercholesterolemia. Am J Cardiol. 1987;59:705–7.
21. Grossman M, Raper SE, Kozarsky K et al. Successful *ex vivo* gene therapy directed to liver in a patient with familial hypercholesterolaemia. Nat Genet. 1994;6:335–41.
22. Grossman M, Rader DJ, Muller DWM et al. A pilot study of *ex vivo* gene therapy for homozygous familial hypercholesterolaemia. Nat Med. 1995;1:1148–54.
23. Fox IJ, Chowdhury JR, Kaufman SS et al. Treatment of the Crigler–Najjar syndrome type I with hepatocyte transplantation. N Engl J Med. 1998;338:1422–6.
24. Muraca M, Gerunda G, Neri D et al. Hepatocyte transplantation as a treatment for glycogen storage disease type 1a. Lancet. 2002;359:317–18.
25. Helene C. Rational design of sequence-specific oncogene inhibitors based on antisense and antigene oligonucleotides. Eur J Cancer. 1991;27:1466–71.
26. Helene C, Toulme JJ. Specific regulation of gene expression by antisense, sense and antigene nucleic acids. Biochim Biophys Acta. 1990;1049:99–125.
27. Haseloff J, Gerlach WL. Simple RNA enzymes with new and highly specific endoribonuclease activities. Nature. 1988;334:585–91.
28. Thompson JD, Macejak D, Couture L, Stinchcomb DT. Ribozymes in gene therapy. Nat Med. 1995;1:277–8.
29. Dykxhoorn DM, Lieberman J. The silent revolution: RNA interference as basic biology, research tool, and therapeutic. Annu Rev Med. 2005;56:401–23.
30. Williams BR. Targeting specific cell types with silencing RNA. N Engl J Med. 2005;353: 1410–11.
31. Kitabwalla M, Ruprecht RM. RNA interference – a new weapon against HIV and beyond. N Engl J Med. 2002;347:1364–7.
32. Taylor JA, Naoumov NV. The potential of RNA interference as a tool in the management of viral hepatitis. J Hepatol. 2005;42:139–44.
33. Calabretta B. Inhibition of protooncogene expression by antisense oligodeoxynucleotides: biological and therapeutic implications. Cancer Res. 1991;51:4505–10.
34. Tseng BY, Brown KD. Antisense oligonucleotide technology in the development of cancer therapeutics. Cancer Gene Ther. 1994;1:65–71.
35. Herschkowitz I. Functional inactivation of genes by dominant negative mutations. Nature. 1987;329:219–22.
36. von Weizsäcker F, Blum HE, Wands JR. Cleavage of hepatitis B virus RNA by three ribozymes transcribed from a single DNA template. Biochem Biophys Res Commun. 1992; 189:743–8.

37. Beck J, Nassal M. Efficient hammerhead ribozyme-mediated cleavage of the structured hepatitis B virus encapsidation signal *in vitro* and in cell extracts, but not in intact cells. Nucl Acids Res. 1995;23:4954–62.
38. Lieber A, He C-Y, Polyak SJ, Gretch DR, Barr D, Kay MA. Elimination of hepatitis C virus RNA in infected human hepatocytes by adenovirus-mediated expression of ribozymes. J Virol. 1996;70:8782–91.
39. Sakamoto N, Wu CH, Wu GY. Intracellular cleavage of hepatitis C virus RNA and inhibition of viral protein translation by hammerhead ribozymes. J Clin Invest. 1996;98: 2720–8.
40. Welch PJ, Tritz R, Yei S, Leavitt M, Yu M, Barber J. A potential therapeutic application of hairpin ribozymes – *in vitro* and *in vivo* studies of gene therapy for hepatitis C virus infection. Gene Ther. 1996;3:994–1001.
41. Shlomai A, Shaul Y. Inhibition of hepatitis B virus expression and replication by RNA interference. Hepatology 2003;37:764-70.
42. Ying C, De Clercq E, Neyts J. Selective inhibition of hepatitis B virus replication by RNA interference. Biochem Biophys Res Commun. 2003;309:482–4.
43. Konishi M, Wu CH, Wu GY. Inhibition of HBV replication by siRNA in a stable HBV-producing cell line. Hepatology. 2003;38:842–50.
44. Giladi H, Ketzinel-Gilad M, Rivkin L, Felig Y, Nussbaum O, Galun E. Small interfering RNA inhibits hepatitis B virus replication in mice. Mol Ther. 2003;8:769–76.
45. McCaffrey AP, Nakai H, Pandey K et al. Inhibition of hepatitis B virus in mice by RNA interference. Nat Biotechnol. 2003;21:639–44.
46. Klein C, Bock CT, Wedemeyer H et al. Inhibition of hepatitis B virus replication *in vivo* by nucleoside analogues and siRNA. Gastroenterology. 2003;125:9–18.
47. Uprichard SL, Boyd B, Althage A, Chisari FV. Clearance of hepatitis B virus from the liver of transgenic mice by short hairpin RNAs. Proc Natl Acad Sci USA. 2005;102:773–8.
48. Grimm D, Streetz KL, Jopling CL et al. Fatality in mice due to oversaturation of cellular microRNA/short hairpin RNA pathways. Nature. 2006;441:537–41.
49. Randall G, Grakoui A, Rice CM. Clearance of replicating hepatitis C virus replicon RNAs in cell culture by small interfering RNAs. Proc Natl Acad Sci USA. 2003;100:235–40.
50. Kapadia SB, Brideau-Andersen A, Chisari FV. Interference of hepatitis C virus RNA replication by short interfering RNAs. Proc Natl Acad Sci USA. 2003;100:2014–18.
51. Wilson JA, Jayasena S, Khvorova A et al. RNA interference blocks gene expression and RNA synthesis from hepatitis C replicons propagated in human liver cells. Proc Natl Acad Sci USA. 2003;100:2783–8.
52. Wagner RW. Gene inhibition using antisense oligodeoxynucleotides. Nature. 1994;372: 333–5.
53. Branch AD. A hitchhiker's guide to antisense and nonantisense biochemical pathways. Hepatology. 1996;24:1517–29.
54. Dougherty WG, Parks TD. Transgenes and gene suppression: telling us something new? Curr Opin Cell Biol. 1995;7:399–405.
55. Goodarzi G, Gross SC, Tewari A, Watabe K. Antisense oligodeoxyribonucleotides inhibit the expression of the gene for hepatitis B virus surface antigen. J Gen Virol. 1990;71:3021–5.
56. Blum HE, Galun E, Weizsäcker F, Wands JR. Inhibition of hepatitis B virus by antisense oligodeoxynucleotides. Lancet. 1991;337:1230.
57. Wu GY, Wu CH. Specific inhibition of hepatitis B viral gene expression *in vitro* by targeted antisense oligonucleotides. J Biol Chem. 1992;267:12436–9.
58. Nakazono K, Ito Y, Wu CH, Wu GY. Inhibition of hepatitis B virus replication by targeted pretreatment of complexed antisense DNA *in vitro*. Hepatology. 1996;23:1297–303.
59. Wakita T, Wands JR. Specific inhibition of hepatitis C virus expression by antisense oligodeoxynucleotides. *In vitro* model for selection of target sequence. J Biol Chem. 1994; 269:14205–10.
60. Mizutani T, Kato N, Hirota M, Sugiyama K, Murakami A, Shimotohno K. Inhibition of hepatitis C virus replication by antisense oligonucleotide in culture cells. Biochem Biophys Res Commun. 1995;212:906–11.
61. Alt M, Renz R, Hofschneider PH, Paumgartner G, Caselmann WH. Specific inhibition of hepatitis C viral gene expression by antisense phosphorothioate oligodeoxynucleotides. Hepatology. 1995;22:707–17.

62. Seki M, Honda Y. Phosphorothioate antisense oligodeoxynucleotides capable of inhibiting hepatitis C virus gene expression: *in vitro* translation assay. J Biochem. 1995;118:1199–204.
63. Hanecak R, Brown-Driver V, Fox MC et al. Antisense oligonucleotide inhibition of hepatitis C virus gene expression in transformed hepatocytes. J Virol. 1996;70:5203–12.
64. Vidalin O, Major ME, Rayner B, Imbach JL, Trepo C, Inchauspe G. *In vitro* inhibition of hepatitis C virus gene expression by chemically modified antisense oligodeoxynucleotides. Antimicrob Agents Chemother. 1996;40:2337–44.
65. Yao Z, Zhou Y, Feng X, Chen C, Guo J. *In vivo* inhibition of hepatitis B viral gene expression by antisense phosphorothioate oligodeoxynucleotides in athymic nude mice. J Viral Hepat. 1996;3:19–22.
66. Offensperger WB, Offensperger S, Walter E et al. *In vivo* inhibition of duck hepatitis B virus replication and gene expression by phosphorothioate modified antisense oligodeoxynucleotides. EMBO J. 1993;12:1257–62.
67. Bartholomew RM, Carmichael EP, Findeis MA, Wu CH, Wu GY. Targeted delivery of antisense DNA in woodchuck hepatitis virus-infected woodchucks. J Viral Hepat. 1995;2: 273–8.
68. Tuerk C, Gold L. Systematic evolution of ligands by exponential enrichment: RNA ligands to bacteriophage T4 DNA polymerase. Science. 1990;249:505–10.
69. Baltimore D. Intracellular immunization. Nature. 1988;325:395–6.
70. Delaney MA, Goyal S, Seeger C. Design of modified core genes that inhibit replication of woodchuck hepatitis virus. In: Hollinger FB, Lemon SM, Margolis H, eds. Viral Hepatitis and Liver Disease. Baltimore, Maryland: Williams & Wilkins, 1991:667–8.
71. Scaglioni PP, Melegari M, Wands JR. Characterization of hepatitis B virus core mutants that inhibit viral replication. Virology. 1994;205:112–20.
72. von Weizsäcker F, Wieland S, Blum HE. Inhibition of viral replication by genetically engineered mutants of the duck hepatitis B virus core protein. Hepatology. 1996;24:294–9.
73. Scaglioni P, Melegari M, Takahashi M, Chowdhury JR, Wands J. Use of dominant negative mutants of the hepadnaviral core protein as antiviral agents. Hepatology. 1996;24:1010–17.
74. Scaglioni PP, Melegari M, Wands JR. Posttranscriptional regulation of hepatitis B virus replication by the precore protein. J Virol. 1997;71:345–53.
75. Buckwold VE, Xu Z, Chen M, Yen TSB, Ou J-H. Effects of a naturally occurring mutation in the hepatitis B virus basal core promoter on precore gene expression and viral replication. J Virol. 1996;70:5845–51.
76. Guidotti LG, Matzke B, Pasquinelli C, Shoenberger JM, Rogler CE, Chisari FV. The hepatitis B virus (HBV) precore protein inhibits HBV replication in transgenic mice. J Virol. 1996;70:7056–61.
77. Blau HM, Springer ML. Muscle mediated gene therapy. N Engl J Med. 1995;333:1554–6.
78. McDonnell WM, Askari FK. DNA vaccines. N Engl J Med. 1996;334:42–5.
79. Pardoll DM, Beckerleg AM. Exposing the immunology of naked DNA vaccines. Immunity. 1995;3:165–9.
80. Schirmbeck R, Bohm W, Ando K, Chisari FV, Reimann J. Nucleic acid vaccination primes hepatitis B virus surface antigen specific cytotoxic T lymphocytes in nonresponder mice. J Virol. 1995;69:5929–34.
81. Vitiello A, Ishioka G, Grey HM et al. Development of a lipopeptide based therapeutic vaccine to treat chronic HBV infection. I. Induction of a primary cytotoxic T lymphocyte response in humans. J Clin Invest. 1995;95:341–9.
82. Kuhöber A, Pudollek HP, Reifenberg K et al. DNA immunization induces antibody and cytotoxic T cell responses to hepatitis B core antigen in H-2b mice. J Immunol. 1996;156: 3687–95.
83. Mancini M, Hadchouel M, Davis HL, Whalen RG, Tiollais P, Michel M-L. DNA-mediated immunization in a transgenic mouse model of the hepatitis B surface antigen chronic carrier state. Proc Natl Acad Sci USA. 1996;93:12496–501.
84. Davis HL, Schirmbeck R, Reimann J, Whalen RG. DNA-mediated immunization in mice induces a potent MHC class I-restricted cytotoxic T lymphocyte response to the hepatitis B envelope protein. Human Gene Ther. 1995;6:1447–56.
85. Lagging LM, Meyer K, Hoft D, Houghton M, Belshe RB, Ray R. Immune responses to plasmid DNA encoding the hepatitis C virus core protein. J Virol. 1995;69:5859–63.

86. Major ME, Vitvitski L, Mink MA et al. DNA based immunization with chimeric vectors for the induction of immune responses against the hepatitis C virus nucleocapsid. J Virol. 1995;69:5798–805.
87. Chow YH, Huang WL, Chi WK, Chu YD, Tao MH. Improvement of hepatitis B virus DNA vaccines by plasmids coexpressing hepatitis B surface antigen and interleukin-2. J Virol. 1997;71:169–78.
88. Wands JR, Lavaissiere L, Moradpour D et al. Immunological approach to hepatocellular carcinoma. J Viral Hepat. 1997;4(Suppl. 2):60–74.
89. Grimm CF, Ortmann D, Mohr L et al. Mouse alpha-fetoprotein-specific DNA-based immunotherapy of hepatocellular carcinoma leads to tumor regression in mice. Gastroenterology. 2000;119:1104–12.
90. Geissler M, Mohr L, Weth R et al. Immunotherapy directed against alpha-fetoprotein results in autoimmune liver disease during liver regeneration in mice. Gastroenterology. 2001;121:931–9.
91. Ido A, Nakata K, Kato Y et al. Gene therapy for hepatoma cells using a retrovirus vector carrying herpes simplex virus thymidine kinase gene under the control of human alpha-fetoprotein gene promoter. Cancer Res. 1995;55:3105–9.
92. Kaneko S, Hallenbeck P, Kotani T et al. Adenovirus-mediated gene therapy of hepatocellular carcinoma using cancer-specific gene expression. Cancer Res. 1995;55: 5283–7.
93. Qian C, Bilbao R, Bruna O, Prieto J. Induction of sensitivity to ganciclovir in human hepatocellular carcinoma cells by adenovirus mediated gene transfer of herpes simplex virus thymidine kinase. Hepatology. 1995;22:118–23.
94. Wills KN, Huang WM, Harris MP, Machemer T, Maneval DC, Gregory RJ. Gene therapy for hepatocellular carcinoma: chemosensitivity conferred by adenovirus-mediated transfer of the HSV-1 thymidine kinase gene. Cancer Gene Ther. 1995;2:191–7.
95. Kanai F, Shiratori Y, Yoshida Y et al. Gene therapy for alpha-fetoprotein-producing human hepatoma cells by adenovirus-mediated transfer of the herpes simplex virus thymidine kinase gene. Hepatology. 1996;23:1359–68.
96. Kanai F, Lan KH, Shiratori Y et al. *In vivo* gene therapy for alpha-fetoprotein-producing hepatocellular carcinoma by adenovirus-mediated transfer of cytosine deaminase gene. Cancer Res. 1997;57:461–5.
97. Mohr L, Shankara S, Yoon SK et al. Gene therapy of hepatocellular carcinoma *in vitro* and *in vivo* in nude mice by adenoviral transfer of the *Escherichia coli* purine nucleoside phosphorylase gene. Hepatology. 2000;31:606–14.
98. Krohne TU, Shankara S, Geissler M et al. Mechanisms of cell death induced by suicide genes encoding purine nucleoside phosphorylase and thymidine kinase in human hepatocellular carcinoma cells *in vitro*. Hepatology. 2001;34:511–18.
99. Mohr L, Rainov NG, Mohr UG, Wands JR. Rabbit cytochrome P450 4B1: a novel prodrug activating gene for pharmacogene therapy of hepatocellular carcinoma. Cancer Gene Ther. 2000;7:1008–14.
100. Kuriyama S, Nakatani T, Masui K et al. Bystander effect caused by suicide gene expression indicates the feasibility of gene therapy for hepatocellular carcinoma. Hepatology. 1995;22: 1838–46.
101. Butterfield LH. Immunotherapeutic strategies for hepatocellular carcinoma. Gastroenterology. 2004;127:S232–41.
102. Butterfield LH, Meng WS, Koh A et al. T cell responses to HLA-A*0201-restricted peptides derived from human alpha fetoprotein. J Immunol. 2001;166:5300–8.
103. Iwashita Y, Tahara K, Goto S et al. A phase I study of autologous dendritic cell-based immunotherapy for patients with unresectable primary liver cancer. Cancer Immunol Immunother. 2003;52:155–61.
104. Cao G, Kuriyama S, Du P et al. Complete regression of established murine hepatocellular carcinoma by *in vivo* tumor necrosis factor alpha gene transfer [see comments]. Gastroenterology. 1997;112:501–10.
105. Huang H, Chen SH, Kosai K, Finegold MJ, Woo SL. Gene therapy for hepatocellular carcinoma: long-term remission of primary and metastatic tumors in mice by interleukin-2 gene therapy *in vivo*. Gene Ther. 1996;3:980–7.
106. Barajas M, Mazzolini G, Genove G et al. Gene therapy of orthotopic hepatocellular carcinoma in rats using adenovirus coding for interleukin 12. Hepatology. 2001;33:52–61.

107. Kroger A, Ortmann D, Krohne TU et al. Growth suppression of the hepatocellular carcinoma cell line Hepa1-6 by an activatable interferon regulatory factor-1 in mice. Cancer Res. 2001;61:2609–17.
108. Sangro B, Mazzolini G, Ruiz J et al. Phase I trial of intratumoral injection of an adenovirus encoding interleukin-12 for advanced digestive tumors. J Clin Oncol. 2004;22:1389–97.
109. Ishikawa H, Nakao K, Matsumoto K et al. Antiangiogenic gene therapy for hepatocellular carcinoma using angiostatin gene. Hepatology. 2003;37:696–704.
110. Heise C, Sampson-Johannes A, Williams A, McCormick F, von Hoff DD, Kirn DH. ONYX-015, an E1B gene-attenuated adenovirus, causes tumor-specific cytolysis and antitumoral efficacy that can be augmented by standard chemotherapeutic agents. Nat Med. 1997;3:639–45.
111. Lowe SW. Progress of the smart bomb cancer virus. Nat Med. 1997;3:606–8.
112. Habib N, Salama H, Abd El Latif Abu Median A et al. Clinical trial of E1B-deleted adenovirus (dl1520) gene therapy for hepatocellular carcinoma. Cancer Gene Ther. 2002;9: 254–9.
113. Pei Z, Chu L, Zou W et al. An oncolytic adenoviral vector of Smac increases antitumor activity of TRAIL against HCC in human cells and in mice. Hepatology. 2004;39:1371–81.
114. Hallenbeck PL, Chang YN, Hay C et al. A novel tumor-specific replication-restricted adenoviral vector for gene therapy of hepatocellular carcinoma. Hum Gene Ther. 1999;10: 1721–33.
115. Pawlik TM, Nakamura H, Yoon SS et al. Oncolysis of diffuse hepatocellular carcinoma by intravascular administration of a replication-competent, genetically engineered herpesvirus. Cancer Res. 2000;60:2790–5.
116. Liu BL, Robinson M, Han ZQ et al. ICP34.5 deleted herpes simplex virus with enhanced oncolytic, immune stimulating, and anti-tumour properties. Gene Ther. 2003;10:292–303.
117. Davis HL, Michel ML, Whalen RG. DNA-based immunization induces continuous secretion of hepatitis B surface antigen and high levels of circulating antibody. Hum Mol Genet. 1993;2:1847–51.
118. Davis HL, McCluskie MJ, Gerin JL, Purcell RH. DNA vaccine for hepatitis B: Evidence for immunogenicity in chimpanzees and comparison with other vaccines. Proc Natl Acad Sci USA. 1996;93:7213–18.
119. Davis HL, Mancini M, Michel ML, Whalen RG. DNA-mediated immunization to hepatitis B surface antigen – longevity of primary response and effect of boost. Vaccine 1996;14:910–15.
120. Michel ML, Davis HL, Schleef M, Mancini M, Tiollais P, Whalen RG. DNA-mediated immunization to the hepatitis B surface antigen in mice: aspects of the humoral response mimic hepatitis B viral infection in humans. Proc Natl Acad Sci USA. 1995;92:5307–11.
121. Forns X, Payette PJ, Ma X et al. Vaccination of chimpanzees with plasmid DNA encoding the hepatitis C virus (HCV) envelope E2 protein modified the infection after challenge with homologous monoclonal HCV. Hepatology. 2000;32:618–25.
122. Folgori A, Capone S, Ruggeri L et al. A T-cell HCV vaccine eliciting effective immunity against heterologous virus challenge in chimpanzees. Nat Med. 2006;12:190–7.

16
Cancer gene therapy: the experience from clinical trials

J. PRIETO, C. QIAN and B. SANGRO

INTRODUCTION

Gene therapy is based on the introduction of genes into cells to control disease. The entry and expression of genes into cells is facilitated by incorporation of the therapeutic genetic material into molecular constructs named gene therapy vectors. In most cases the vectors consist of viruses that are deleted of a part of their genome which is substituted by the therapeutic gene (transgene) and appropriate regulatory elements. Viral vectors differ in their biological properties including cloning capacity and ability to express the transgene for short or long periods of time. Some viral vectors (adenovirus) remain episomal, while others integrate into the host genome with high (lentivirus, SV40) or low (adeno-associated virus, AAV) efficiency. First-generation adenoviruses enable only short duration of the expression of the transgene due to strong immune response against the viral proteins expressed by the infected cells. By contrast, third-generation adenoviruses (also called gutless adenoviruses), which are devoid of all adenoviral genes, allow long-term transgene expression.

Gene therapy is a highly plastic procedure that can be used in many different ways to combat cancer. The aim of the intervention could be: (a) to induce a direct lysis of the tumour cells, (b) to stimulate antitumour immunity, (c) to block tumour growth by changing the biological conditions of the tumour environment, (d) to combine various of these effects. Because of difficulties in transducing the tumour by the intravenous route, direct injection of the vector into the tumour nodule or in the peritumoural tissue (with echographic guidance) has frequently been used to treat patients with cancer.

TREATMENT MODALITIES

Up to now the following antitumour strategies have been tested in gene therapy clinical trials: (a) infection of the tumour with viral vectors encoding suicide genes such as thymidine kinase (HSV-tk), which converts ganciclovir into a toxic compound that kills the infected cells, (b) transference to the neoplastic cells of the tumour-suppressor gene p53, (c) the use of viruses with the ability to selectively replicate in tumour cells such as ONYX-015 (an adenovirus vector deleted of the viral gene E1B), or (d) transduction of the tumour or peritumoural tissue with vectors encoding cytokines with immunostimulatory activity such as IL-12.

Suicide genes

Genes encoding for enzymes that convert an innocuous prodrug to a toxic compound are known as suicide genes. Thymidine kinase (tk) from HSV-1 is the best-characterized prodrug activation enzyme[1]. It converts ganciclovir (a well-tolerated antiviral drug) into a toxic phosphorylated compound that inhibits both nuclear and mitochondrial DNA synthesis leading to cell death[1,2]. This strategy permits concentration of a toxic metabolite in the tumoural tissue avoiding systemic toxicity. An interesting property of suicide genes is the so-called bystander effect that results from the diffusion of the activated prodrug to non-transduced neighbouring cells thus enhancing the antitumour activity of the procedure[3]. The bystander effect may also derive from the stimulation of the antitumoral immune response as result of necrosis and apoptosis of the neoplastic cells[4].

In a recent trial[5] a first-generation adenovirus encoding HSV-tk under the control of CMV promoter (AdCMVtk) was used to treat seven patients with hepatocellular carcinoma (HCC) by intratumoral administration of the vector in a dose-escalation fashion from 2×10^{10} to 2×10^{12} viral particles (v.p.) in consecutive patients. Contrary to patients treated with lower doses, those who received 10^{12} v.p. or more ($n = 4$) showed stabilization of the treated tumour and in two of them wide areas of necrosis were apparent in the CT scan performed at day 30 after therapy. Tolerance was good and no significant side-effects were apparent, but benefit was transient. In this trial positron emission tomography (PET) imaging was employed to monitor transgene expression using a labelled substrate of HSV-tk ([^{18}F]FHBG). With this methodology strong retention of the label was found in the treated tumour nodules in all patients who received 10^{12} v.p. or more, but not in those given lower vector doses. Interestingly the surrounding non-tumoral cirrhotic tissue was completely spared from transduction in all cases. The restriction of tissue transduction to the injected tumour nodule argues in favour of the safety of procedure when applied to patients with liver cancer implanted on a cirrhotic liver. This study demonstrated that human HCC are easily infected with adenoviral vectors but it also showed that a second dose of the vector failed to transduce the tumour, possibly as result of the development of neutralizing anti-adenoviral antibodies.

An antitumour effect against HCC and metastatic liver cancer was observed using other prodrug activating enzymes including: (a) *Escherichia coli* purine nucleoside phosphorylase that converts fludarabine into a diffusible toxic compound[6], (b) cytosine deaminase that turns antifungal 5-fluocytosine (5-FC) into 5-fluoracyl[7,8], and (c) bacterial nitroreductase which converts CB1954 (5-(azidrin-1-yl)-2,4-dinitrobenzamide) to a short-lived, very toxic DNA crosslinking compound. Recently, a clinical trial in patients with primary and metastatic liver cancer using intratumoral injection of escalating doses (10^8–5×10^{11} v.p.) of a first adenovirus encoding nitroreductase showed good tolerance of the procedure and a dose-related expression of the transgene in the biopsy of the treated tumour[9].

Tumour-suppressor genes

It has been reported that the transference of wild-type p53 gene to p53-negative HCC cells inhibits tumour growth and increases the sensitivity to chemotherapy[10]. However, only a modest therapeutic effect was obtained in clinical trials employing adenovirus encoding p53[11]. The limited efficacy of p53 gene supplementation reflects the inability to transduce efficiently all the tumour cells with vectors presently available, and also the impossibility of transducing the tumour repeatedly due to the generation of neutralizing antibodies after the first dose of the vector.

Oncolytic virotherapy

This strategy is based on the use of viruses that replicate and kill preferentially tumour cells. These cells then become the source of new viral particles that infect surrounding tumour cells. Adenoviruses have been extensively used for the production of oncolytic agents. The first of this kind of virus (named dl1520 or ONYX-015) is based on the deletion of the viral gene E1B 55K, which makes its replication dependent on the defect of the p53 pathway in the infected cells[12]. In clinical trials ONYX-015 was safely administered by intratumoral injection[13] and in a pilot study in patients with primary and metastatic liver cancer the therapy was well tolerated and some clinical responses were observed[14]. However, a further study in which repeated intratumoral injections of this replication-selective adenovirus were given to patients with HCC failed to show clinically relevant efficacy[15]. In another trial, in subjects with hepatobiliary tumours, intralesional administration of the vector provoked partial responses in less than 10% of the patients[16]. The poor efficacy of oncolytic adenoviruses may be related to the generation of a strong humoral and cellular immune response against adenoviral antigens that limits the spread of the virus and blocks the infectivity of subsequent doses of the vector.

Immunogene therapy

Tumours express neoantigens that could elicit protective immunity but antitumour immune responses are in general weak or ineffective due to low expression of MHC molecules[17], production of immunosuppressive factors (TGF-β, IL-10, VEGF, or IL-8)[18], induction of $CD4^+$ $CD25^+$ regulatory T cells (Treg)[19] or expression of Fas ligand by neoplastic cells resulting in apoptosis of tumour-infiltrating T cells[20]. On the other hand tumour cells can exploit host-derived cytokines to increase resistance to apoptosis and to stimulate growth and dissemination[21]. Gene therapy can be used to overcome these barriers and to promote antitumour responses. Cytokines (such as IL-2, IL-12, IL-15, TNF-α and IFN-γ) generated during the immune response, in inflammation and in infection, can restrict tumour growth, and it has been shown that their systemic administration could elicit antitumour effects[21]. However, the severe toxicity observed when these cytokines are given by parenteral injection limits their application.

Gene transfer of immunostimulatory cytokines to the neoplastic cells leads to high local concentration with low blood concentration and reduced systemic toxicity. A variety of immunostimulatory cytokines, chemokines or co-stimulatory molecules have been employed with efficacy in gene therapy of diverse tumour models[22]. IL-12 is a particularly potent antitumour cytokine which induces a Th1 type of response[23,24], activates cytotoxic T lymphocytes and natural killer (NK) cells. The antitumour activity of IL-12 also stems from its robust anti-angiogenic properties[25–27]. In fact, of all immunogene therapeutic strategies used in experimental models of cancer only IL-12-based gene therapy has been transferred to the clinic for the treatment of patients with liver cancer[28]. The clinical trial was performed in patients with liver tumours either primary or secondary to colorectal or pancreatic carcinomas. Each subject received one to three intratumoral injections of a first-generation adenoviral vector encoding human IL-12. The treatment was without any significant side-effects and the maximal tolerated dose was not reached. However, the antitumour effect was weak, with only one partial response in a patient with HCC. In general patients with HCC had a better outcome than other histological groups. These modest antitumour effects were probably due to the low and short-lived expression of the transgene when using first-generation adenovirus, and also to the development of neutralizing antibodies which abrogated tumour transduction after repeated vector doses.

Engineered dendritic cells (DC)

Genetic manipulation of DC is a promising approach to activate antitumour immunity. It has been shown that intratumoral injection of DC engineered to produce IL-12 can induce tumour regression in animal models of digestive cancer[29]. A clinical trial was recently performed in patients with primary or metastatic liver cancer using intratumoral injections of monocyte-derived DC transduced with first-generation adenoviral vectors encoding IL-12[30]. The patients received up to three equal doses of DC (doses were escalated from 10 to 50 millions of cells in three cohorts of patients) at 21-day intervals. The expectation was that the modified DC would take up tumour antigens and migrate to lymph nodes to activate a specific Th1 antitumoral response by locally secreting IL-12. However, although the procedure was well tolerated, the elicited antitumour effect was weak. By labelling the DC with a radioactive compound it was found that the tumour retained the injected DC and prevented their migration to regional lymph nodes, rendering the therapy inefficient. The retention of DC in the tumour might be due to the generation of high amounts of IL-8 by the neoplastic tissue[31]. Results from this trial indicate that intratumour injection of DC is inefficient at inducing potent antitumour immunity and it should be avoided. Instead mature DC, genetically engineered or not (once pulsed with tumour lysate or tumour

RNA), should be injected inside or in the vicinity of lymph nodes. This form of therapy might be useful to eliminate minimal residual disease after surgery rather than to control growth of advanced cancer.

PERSPECTIVES OF GENE THERAPY OF LIVER CANCER

Although gene therapy is still in its infancy, clinical trials have shown that it represents a feasible and promising approach to treat cancer. Up to now, first-generation adenoviruses have been the type of vector most frequently used for this purpose. From gene thereapy clinical trials we have learnt that these vectors transduce liver tumours efficiently, but the expression of the transgene is short-lived and repeated transduction of the neoplasm is not feasible due to the generation of neutralizing anti-adenoviral antibodies after the first vector dose. It is expected that long-term expression vectors would enhance the effectiveness of cancer gene therapy[32]. This type of vector (equipped with appropriate inducible promoters) would enable prolonged and regulable expression of the therapeutic transgene (or combination of transgenes) inside or around the neoplastic lesions facilitating antitumour responses. Helper-dependent adenoviruses (also called gutless adenovirus) have emerged as a promising long-term expression vector to treat malignancies since, in addition to allowing durable transgene expression, they also have very high cloning capacity which permits allocation of different transgenes with different promoters in the same vector[33]. It is thus possible to induce the production of diverse antitumour molecules and to control the expression of each one with an independent inducible promoter using a single vector. In this manner we can direct the antitumour attack to different biological targets, making the therapy more effectual.

In addition to improvements in vector technology, with the progress in the knowledge of tumour biology novel targets have been identified. This creates new opportunities for gene therapy which can rely not only on immunogene strategies, oncolytic virotherapy or suicide genes, but also on the expression inside the tumour mass of humanized monoclonal antibodies or decoy molecules directed to block factors that are crucial for tumour progression such as growth factors, growth factor receptors, metalloproteases or integrins. In other words, gene therapy can be used to produce, within the tumour mass itself, a variety of molecules that can disrupt the biological microenvironment that the tumour needs to grow.

Moreover, new developments in stem cell biology have opened new horizons to fight cancer. In this regard, the recruitment of circulating progenitor cells by the tumour can be exploited by genetic engineering of these cells that can serve as a vehicle of therapeutic genes. A theoretical advantage of combining cell and gene therapy is that it would be possible to target all metastatic tumour lesions after the intravenous injection of the engineered progenitors.

References

1. Fillat C, Carrio M, Cascante A, Sangro B. Suicide gene therapy mediated by the Herpes Simplex virus thymidine kinase gene/ganciclovir system: fifteen years of application. Curr Gene Ther. 2003;3:13–26.
2. Herraiz M, Beraza N, Solano A et al. Liver failure caused by herpes simplex virus thymidine kinase plus ganciclovir therapy is associated with mitochondrial dysfunction and mitochondrial DNA depletion. Hum Gene Ther. 2003;14:463–72.
3. Mesnil M, Yamasaki H. Bystander effect in herpes simplex virus–thymidine kinase/ ganciclovir cancer gene therapy: role of gap-junctional intercellular communication. Cancer Res. 2000;60:3989–99.
4. Kianmanesh AR, Perrin H, Panis Y et al. A 'distant' bystander effect of suicide gene therapy: regression of nontransduced tumors together with a distant transduced tumor. Hum Gene Ther. 1997;8:1807–14.
5. Peñuelas I, Mazzolini G, Boán JF et al. Positron emission tomography imaging of adenoviral-mediated transgene expression in liver cancer patients. Gastroenterology. 2005;128:1787–96.
6. Mohr L, Shankara S, Yoon SK et al. Gene therapy of hepatocellular carcinoma *in vitro* and *in vivo* in nude mice by adenoviral transfer of the *Escherichia coli* purine nucleoside phosphorylase gene. Hepatology. 2000;31:606–14.
7. Ohwada A, Hirschowitz EA, Crystal RG. Regional delivery of an adenovirus vector containing the *Escherichia coli* cytosine deaminase gene to provide local activation of 5-fluorocytosine to suppress the growth of colon carcinoma metastatic to liver. Hum Gene Ther. 1996;7:1567–76.
8. Humphreys MJ, Ghaneh P, Greenhalf W et al. Hepatic intra-arterial delivery of a retroviral vector expressing the cytosine deaminase gene, controlled by the CEA promoter and intraperitoneal treatment with 5-fluorocytosine suppresses growth of colorectal liver metastases. Gene Ther. 2001;8:1241–7.
9. Palmer DH, Mautner V, Mirza D et al. Virus-directed enzyme prodrug therapy: intratumoral administration of a replication-deficient adenovirus encoding nitroreductase to patients with resectable liver cancer. J Clin Oncol. 2004;22:1546–52.
10. Xu GW, Sun ZT, Forrester K, Wang XW, Coursen J, Harris CC. Tissue-specific growth suppression and chemosensitivity promotion in human hepatocellular carcinoma cells by retroviral-mediated transfer of the wild-type p53 gene. Hepatology. 1996;24:1264–8.
11. Warren RS, Kirn DH. Liver-directed viral therapy for cancer p53-targeted adenoviruses and beyond. Surg Oncol Clin N Am. 2002;11:571–88.
12. McCormick F. Cancer-specific viruses and the development of ONYX-015. Cancer Biol Ther. 2003;2:S157–60.
13. Kirn D. Oncolytic virotherapy for cancer with the adenovirus dl1520 (Onyx-015): results of phase I and II trials. Expert Opin Biol Ther. 2001;1:525–38.
14. Habib NA, Sarraf CE, Mitry RR et al. E1B-deleted adenovirus (dl1520) gene therapy for patients with primary and secondary liver tumors. Hum Gene Ther. 2001;12:219–26.
15. Habib N, Salama H, Abd El Latif Abu Median A et al. Clinical trial of E1B-deleted adenovirus (dl1520) gene therapy for hepatocellular carcinoma. Cancer Gene Ther. 2002;9: 254–9.
16. Makower D, Rozenblit A, Kaufman H et al. Phase II clinical trial of intralesional administration of the oncolytic adenovirus ONYX-015 in patients with hepatobiliary tumors with correlative p53 studies. Clin Cancer Res. 2003;9:693–702.
17. Algarra I, Cabrera T, Garrido F. The HLA crossroad in tumor immunology. Hum Immunol. 2000;61:65–73.
18. Ranges GE, Figari IS, Espevik T, Palladino MA Jr. Inhibition of cytotoxic T cell development by transforming growth factor beta and reversal by recombinant tumor necrosis factor alpha. J Exp Med. 1987;166:991–8.
19. Antony PA, Restifo NP. $CD4^{+}CD25^{+}$ T regulatory cells, immunotherapy of cancer, and interleukin-2. J Immunother. 2005;28:120–8.
20. Strand S, Hofmann WJ, Hug H et al. Lymphocyte apoptosis induced by CD95 (APO-1/ Fas) ligand-expressing tumor cells – a mechanism of immune evasion? Nat Med. 1996;2: 1361–6.

21. Dranoff G. Cytokines in cancer pathogenesis and cancer therapy. Nat Rev Cancer. 2004;4: 11–22.
22. Prieto J, Qian C, Hernandez-Alcoceba R et al. Gene therapy of liver diseases. Expert Opin Biol Ther. 2004;4:1073–91.
23. Trinchieri G. Interleukin-12: a cytokine at the interface of inflammation and immunity. Adv Immunol. 1998;70:83–243.
24. Mazzolini G, Narvaiza I, Martinez-Cruz LA et al. Pancreatic cancer escape variants that evade immunogene therapy through loss of sensitivity to IFNgamma-induced apoptosis. Gene Ther. 2003;10:1067–78.
25. Sgadari C, Angiolillo AL, Tosato G. Inhibition of angiogenesis by interleukin-12 is mediated by the interferon-inducible protein 10. Blood. 1996;87:3877–82.
26. Mazzolini G, Qian C, Narvaiza I et al. Adenoviral gene transfer of interleukin 12 into tumors synergizes with adoptive T cell therapy both at the induction and effector level. Hum Gene Ther. 2000;11:113–25.
27. Barajas M, Mazzolini G, Genove G et al. Gene therapy of orthotopic hepatocellular carcinoma in rats using adenovirus coding for interleukin 12. Hepatology. 2001;33:52–61.
28. Sangro B, Mazzolini G, Ruiz J et al. Phase I trial of intratumoral injection of an adenovirus encoding interleukin-12 for advanced digestive tumors. J Clin Oncol. 2004;22:1389–97.
29. Melero I, Duarte M, Ruiz J et al. Intratumoral injection of bone-marrow derived dendritic cells engineered to produce interleukin-12 induces complete regression of established murine transplantable colon adenocarcinomas. Gene Ther. 1999;6:1779–84.
30. Mazzolini G, Alfaro C, Sangro B et al. Intratumoral injection of dendritic cells engineered to secrete interleukin-12 by recombinant adenovirus in patients with metastatic gastrointestinal carcinomas. J Clin Oncol. 2005;23:999–1010.
31. Feijoo E, Alfaro C, Mazzolini G et al. Dendritic cells delivered inside human carcinomas are sequestered by interleukin-8. Int J Cancer. 2005;116:275–81.
32. Wang L, Hernandez-Alcoceba R, Shankar V et al. Prolonged and inducible transgene expression in the liver using gutless adenovirus: a potential therapy for liver cancer. Gastroenterology. 2004;126:278–89
33. Guan M, Rodriguez-Madoz JR, Alzuguren P et al. Increased efficacy and safety in the treatment of experimental liver cancer with a novel adenovirus–alphavirus hybrid vector Cancer Res. 2006;66:1620–9.

17
Stem cell transplantation

S. GUPTA, K. CHENG and N. FLEISCHER

INTRODUCTION

A variety of genetic and acquired disorders are amenable to cell and gene therapy. Over the past two decades a large body of work in laboratory animals has established that transplanted cells can engraft, function and proliferate in the liver. Moreover, under suitable conditions, transplanted cells can repopulate the liver. Replacement of diseased hepatocytes with healthy transplanted cells leads to the correction of secretory or metabolic deficiencies, as well as heptotoxicity from various mechanisms[1]. Similarly, the use of genetically modified cells offers unique opportunities for developing novel therapies. However, the scarcity of donor organs is a major problem for cell therapy programmes. In this respect the use of stem cells is of obvious interest. In particular, autologous stem cells that could be transplanted without deleterious host immune responses against the major histocompatibility complex are attractive[2]. On the other hand, transplantation of allogeneic cells will also be useful, particularly since rejection of allografted cells can be prevented by immunosuppressive drugs[3]. The use of xenogeneic cells, e.g. from specific-pathogen-free pigs has been proposed, and is being investigated, to remedy organ shortages[4], although to develop suitable insights in the potential of xenogeneic cells requires much experimental work[5].

WHAT IS A STEM CELL, OR DOES IT REALLY MATTER?

Stem cells can be defined in many ways, including specific biological properties, e.g. capacity for self-renewal and pluripotency (ability to originate cells of all germlayers); age-specific origin from embryonic, fetal or adult tissues, etc. These stem cell issues are summarized at various domains, e.g. http://stemcells.nih.gov/ and http://www.isscr.org/science/index.htm. Although knowledge concerning the existence of stem cells is not new, the isolation of mouse embryonic stem cells in 1980, and of human embryonic stem cells in 1998, generated great scientific interest[6], while simultaneously initiating public debate concerning the need, ethics and limits of stem cell research.

A unique aspect of stem cells, e.g. human embryonic stem cells (hESC), is that these cells can proliferate indefinitely without entering replicative senescence *in vitro*. Although pluripotential hESC offer many advantages, limited understanding of how hESC could be differentiated into liver cells without risks for cancer and other problems is a significant hurdle. As culture conditions for maintaining stem cells are only imperfectly developed at present, cells may acquire genetic lesions or other perturbations that may pose problems[7]. Moreover, studies demonstrating the *in-vivo* fate of hESC have generally been limited[8,9]. For instance, it has not been established that hESC-derived cells can in fact proliferate in the same manner, in intact animals, as can mature hepatocytes. Indeed, in response to tissue-specific lineage advancement in hESC, cell proliferation begins to cease.

The availability in recent years of fetal and adult stem cells, from within the liver and outside of it, has also been of interest for clinical applications. Most experts agree that stem/progenitor cells play roles in the liver[10,11], although the definitive hepatic stem cell has neither been isolated nor fully defined. Under specific circumstances, candidate liver stem cells do become activated in the adult liver. Moreover, during embryological liver development, cell–cell interactions between parenchymal cells of endodermal origin and non-parenchymal cells of mesodermal origin, e.g. endothelial cells are critical in organogenesis with interplays among regulatory cytokines or growth factors[12]. On the other hand, mature hepatocytes that exist in the liver in non-proliferative states in G0 are unable to proliferate *in vitro*, whereas after transplantation in the liver of recipients with diseased hepatocytes that are unable to proliferate, adult hepatocytes can proliferate extensively and indefinitely across multiple generations of animals[13]. This dissociation between *in-vitro* and *in-vivo* proliferation capacity of cells has important ramifications, particularly for organ replacement therapies. Therefore, from a cell therapy perspective, it is not unnatural to ask whether stem cells are truly necessary, or whether cells possessing liver-repopulating properties, regardless of the stem/progenitor cell or differentiated cell phenotype, should be of greater interest.

Despite ongoing uncertainties concerning liver-derived stem/progenitor cells, it is now generally accepted that the liver contains stem cells, although the cell compartment where stem cells reside in the adult liver remains under investigation[11,12]. Similarly, so-called oval cells that exhibit progenitor cell properties can be isolated from the adult liver or pancreas, and under suitable circumstances these cells can produce mature hepatocytes, including after transplantation into recipients. However, stem/progenitor cells of any type are thought to be extremely rare in the healthy adult liver. By contrast, the fetal human liver is enriched in stem/progenitor cells[14–17]. As the fetal liver is a major site for extramedullary haematopoiesis, besides hepatic stem/progenitor cells, haemangioblastic stem/progenitor cells capable of generating haematopoietic lineages, as well as other mesenchymal lineages, e.g. endothelial cells, are abundantly present in this organ. Fetal stem/progenitor cells can be readily isolated and banked[15]. Therefore, whether clinical applications should be developed with these cells has been another significant consideration. Recent studies have demonstrated that fetal human liver cells possess significant proliferation capacity, multilineage differentiation potential, liver-repopulation capacity following transplantation into adult animals, as well as the capacity to reconstitute deficient functions and to promote survival in acute liver failure[15,17–24].

Extrahepatic stem cells, e.g. from the bone marrow, umbilical cord blood, amniocentesis fluid or placenta, are of interest because autologous stem cells could be stored for later use and allogeneic cells may also be used. Although the plasticity of haematopoietic stem cells and bone marrow-derived multipotent adult progenitor (mesenchymal) cells has consumed much recent effort, questions concerning various issues are not fully resolved at present. The general agreement is that blood- or bone marrow-derived stem cells can produce occasional mature hepatocytes, as well as other liver cell types, although the efficiency by which hepatocytes may be produced is low, and cell fusion is considered to be problematic.

Therefore, consideration of several types of stem cells will be appropriate for cell and gene therapy applications. Moreover, differentiation of stem cells to obtain a variety of differentiated cell types should be helpful in developing novel cellular systems to address biological issues, e.g. organ development, remodelling, angiogenesis, cell–cell interactions, etc. studies of pathophysiological mechanisms with stem cells from affected individuals, drug development, toxicological analyses, testing of gene therapy strategies, and other translational applications. Similarly, successful repopulation of the animal liver with transplanted human cells will help obtain chimeric animal models for specific diseases, e.g. viral hepatitis.

CENTRAL ASPECTS OF STEM CELL BIOLOGY FROM A TRANSLATIONAL PERSPECTIVE

As indicated above, the ability to obtain mature hepatocytes from stem cells will greatly facilitate studies of physiological and pathobiological mechanisms, development of *in-vitro* and *in-vivo* models, disease-specific diagnostic tools, suitable mechanisms for genetic manipulation, and therapeutic studies. Mechanisms that result in the advancement of pluripotential hESC-derived germlayers and/or tissue-specific stem/progenitor cells during organ development will also be helpful in recapitulating controlled differentiation of stem cells. Nonetheless, transplantation of undifferentiated hESC produces teratomas or teratocarcinomas, although the intermediary step of 'embryoid body' formation before cell dissociation decreases teratogenicity in hESC[25,26]. Various strategies have been applied to select hepatic-like hESC subpopulations, e.g. through the α-fetoprotein (AFP) promoter trap, expression of the endoderm-specifying Foxa (HNF-3β) transcription factor, or use of cell cycle regulatory chemicals, such as sodium butyrate[27–30]. Similarly, soluble signals have been helpful in generating hepatocyte-like cells without hESC embryoid body formation[30,31]. Genetic profiling of undifferentiated and differentiated hESC showed onset of fetal hepatic enriched genes in embryoid bodies[30]. Furthermore, hESC have been shown to be capable of generating suitable liver cell types after transplantation in appropriate animal models[8].

The ability to identify suitable conditions for culturing stem cells or their derivatives without oncogenetic events or cell transformation will be critical for basic and translational studies. Moreover, the availability of sufficient lines of genetically diverse hESC for studies of various mechanisms will be critical. In principle, stem cells should be devoid of animal products if these cells are to be used for transplantation into humans. Therefore, these and other considerations indicate that a far larger number of hESC lines will be necessary than are available at present to investigators. This ought to require societal resolution of debates concerning derivation of new hESC lines, including individual-specific cells for eventual cell therapies. Detailed studies will then be necessary with specific new hESC lines to recapitulate critical mechanisms, including molecular and epigenetic regulation of gene activation and silencing, regulation of DNA replication, cell cycling, symmetric and/or asymmetric division, and senescence, as well as mechanisms of cell–cell interactions, cell signalling and fate specification. Although these and many related areas are currently under active investigation with existing cell lines, this work will probably consume years and decades.

Another issue concerns niches and compartments of endogenous stem/progenitor cells in tissues. This is necessary for understanding how these types of cells serve roles in organ development and homeostasis. Moreover, under certain circumstances, activation of endogenous stem/progenitor cells should be sufficient for promoting organ repair. On the other hand, to replenish deficient stem/progenitor cells, it should be helpful if mechanisms regulating their persistence or loss could be defined.

In general, the nature of cell-specific signals responsible for lineage specification and differentiation are better understood for the embryonic liver of animals rather than of humans, although major gaps remain. For instance, it is unclear what specific cell compartment originates hepatic stem cells. Recent studies of the intact fetal human liver, as well as of epithelial stem/progenitor cells isolated from the fetal human liver, indicate multilineage differentiation potential in cells, with the capacity to generate both mesenchymal and epithelial derivatives[17]. The fetal human liver develops rapidly during the first trimester, such that the acinar structure is acquired within 7 weeks, although the liver plate structure is not fully developed and the liver simultaneously produces haematopoietic cells and epithelial cells at this stage[14]. Most parenchymal fetal epithelial cells express markers of multiple lineages, including hepatic (AFP, albumin, α_1-antitrypsin, α_1-microglobulin, glucose-6-phosphatase, glycogen, etc.), biliary (CK-19, γ-glutamyltranspeptidase, dipeptidyl peptidase IV), mesenchymal (von Willebrand factor, desmin, vimentin, α-smooth muscle actin) and even undifferentiated stem cell (Oct-4, nanog) markers[15–17] (our unpublished data). Fetal liver stem/progenitor cell populations can be isolated by the display of unique cell surface markers and cells exhibit major histocompatibility antigens poorly, which should attenuate host immune responses[22,32,33]. Fetal human liver cells can also be cryopreserved, expanded in cell culture conditions, and are amenable to genetic modification using efficient viral vectors, e.g. lentivirus vectors[15,34]. Nonetheless, fetal liver cells undergo attrition of telomere lengths during culture in association with decline in cell proliferation although, after reconstitution of telomerase activity, fetal cells retain indefinite proliferation capacity[35]. In such stem/progenitor cells in the fetal liver, additional genetic manipulation, e.g. expression of Pdx-1 gene, which is a homeobox regulator of pancreatic organogenesis and β-cell development, resulted in expression of insulin and other pancreatic genes[36,37]. This was in agreement with the concept of stem cell plasticity, and raised much interest in the potential of fetal cells for treating type 1 diabetes mellitus. Similarly, fetal liver cells were discovered to alter the behaviour of hESC under coculture conditions, leading to efficient production of cells in the erythroid lineage[38]. Recently, transplantation studies of rat or mouse fetal liver cells have verified that fetal cells can mature rapidly, produce and secrete albumin, and proliferate extensively in the liver of syngeneic recipients[18–20,22]. Fetal human liver cells also engraft in the liver or other locations in immunodeficient mice[15]. These considerations indicate that fetal liver cells will have suitable potential for clinical applications, and this needs further development.

The location of fetal epithelial stem/progenitor cells within the liver parenchyma compared with the location of putative stem/progenitor cells in the adult human liver, such as the canal of Hering[39], suggests that liver stem cells may be sequestered in ductal niches of the adult liver rather than primarily originating from such locations during embryonic development. The issues of liver stem/progenitor cell origin, maintenance and sequestration will be helpful for pathophysiological studies, while providing additional indications for cell isolation strategies and analysis of the differentiated fate of cells. Moreover, these analyses will provide suitable ways to assess the properties of stem/progenitor cells with robust *in-vitro* and *in-vivo* assays to understand whether organ repopulating capacity is retained or lost in cells following manipulations aimed at altering cell proliferation or differentiation.

Activation of endogenous tissue-resident stem cells to repair organ damage is an attractive concept, although this requires greater insights in relevant mechanisms. In some instances, e.g. after partial hepatectomy or an acute injury, liver repair mechanisms are better understood than in other conditions[10,11]. Nonetheless, removal of toxins or initiators of injury, where possible; minimization of the consequences of injury; and provision of metabolic support, promotes hepatocellular proliferation and hepatic recovery. Of course, many people suffer extensive liver injury, where orthotopic liver transplantation (OLT) becomes necessary. Were it possible to activate proliferation in remaining hepatic cells in the setting of chronic injury, perhaps through recruitment of resident stem/progenitor cells or by reseeding the liver with transplanted cells capable of liver repopulation[40], liver repair should be feasible without OLT.

Recently, much interest has focused on the possibility that haematopoietic stem cells or other types of stem/progenitor cells could be transplanted in humans with chronic liver disease or liver failure. The idea that circulating cells would contribute in organ repair followed studies showing the capacity of haematopoietic stem cells to differentiate into hepatocytes[41–43]. Subsequently, specific subsets of bone marrow-derived haematopoietic stem cells, as well as mesenchymal cells (multipotent adult progenitor cells – MAPC), were shown to be capable of generating multiple cell lineages[44,45]. The umbilical cord blood is another source of stem cells, including haematopoietic and mesenchymal cells, and this has led to routine clinical applications of cord blood cells. Moreover, umbilical cord blood cells possess the capacity to generate hepatocytes in animals[46–50]. However, the frequency by which cord blood cells produce hepatocytes was small, reminiscent of bone marrow-derived haematopoietic stem cells. Similarly, amniotic fluid has been found to contain epithelial cells in animals, as well as in humans[51,52], and a large fraction of amniotic epithelial cells express liver genes, such as albumin and α_1-antitrypsin. Transplantation of placenta-derived epithelial cells in animals results in integration of transplanted cells in the liver parenchyma with secretion of hepatic proteins[53]. The best-studied extrahepatic stem cell fraction is bone marrow-derived cells. In mouse studies the generation of hepatocytes after bone marrow transplantation required some 7 weeks[54]. Single bone marrow-derived mouse haematopoietic stem cells were sufficient for generating multiple lineages besides liver cells[55]. Similarly, transplantation of MAPC into mouse blastocysts resulted in the production of cells in all three germlayers[56], verifying their multipotentiality. Among convincing demonstrations of the hepatic differentiation potential of bone marrow-derived cells are FAH mouse studies, in which transplantation of healthy bone marrow-derived cells resulted in extensive liver repopulation with disease correction[44]. This type of therapeutic potential was a surprise. However, it soon became apparent that conversion of bone marrow-derived cells into hepatocytes was infrequent[54,57–59]. Moreover, bone marrow-derived haematopoietic stem cells, as well as umbilical cord blood cells, were found to fuse with native cells[49,60,61], myelomonocytic cells in particular[61], although in other studies cell fusion was not found[62,63]. As cell fusion could trigger genetic instability, e.g. aneuploidy, this process has raised concerns; therefore, many uncertainties exist with respect to the potential of extrahepatic circulating stem/progenitor cells, including the efficiency by which circulating stem cells will target to the liver, the process by which these cells will generate hepatocytes, and whether transplanted cells will be beneficial not because hepatocytes are produced *de novo* but because cytoprotective or other factors are released by these cells in the hepatic microenvironment. Thus knowledge of the nature and role of soluble signals and transcriptional regulators capable of activating or suppressing the emergence of endogenous stem cells will be helpful.

CELL COMPARTMENT-SPECIFIC CONSIDERATIONS FOR STEM CELL APPLICATIONS

Hepatocytes constitute 60% of all liver cell types and are the single largest cellular component of the liver. Moreover, abnormal hepatocyte function through genetic or acquired causes encompasses a large group of disorders, which are potentially amenable to stem cell transplantation. As liver failure, acute and chronic, is an important indication for cell therapy, it is understandable that early efforts were directed towards this disorder. However, analysis of cell therapy in acute liver failure has been hampered by the deficiency of outstanding animal models capable of reproducing this condition. The general hope is that cell transplantation in acute liver failure will offer critical metabolic support in the short term followed by promotion of regeneration in the native liver. In chronic liver failure the immediate hope is that, by providing additional synthetic or metabolic liver support, cell therapy will ameliorate the complications of coagulopathy, hypoproteinaemia and encephalopathy. Whether cell transplantation will help in regression of liver fibrosis and portal hypertension is less clear.

At present, studies of stem cell transplantation in humans are in the very early stages. Certainly, fetal human liver cells have been transplanted in few patients with acute liver failure, although the precise value of cell therapy in this condition remains rather unclear in view of the limited experience[23,24]. For instance, in seven patients with acute liver failure and grade III to IV encephalopathy, fetal human hepatocytes were transplanted intraperitoneally[23]. Four patients died after cell transplantation, whereas three recovered. Subsequently, another patient with acute liver failure due to fatty liver of pregnancy recovered following transplantation of fetal human hepatocytes in the peritoneal cavity[24]. An advantage of using the peritoneal cavity for this purpose is that large numbers of cells can be transplanted with the expectation that these cells will provide metabolic support.

Similarly, to improve recovery in patients with chronic liver disease and complications of liver failure, transplantation of relatively large numbers of cells will be necessary to create an extrahepatic reservoir of liver function. Whether bioengineering principles can be combined with cell transplantation to accomplish neo-organogenesis is being studied. On the other hand, to aid recovery in liver failure will probably require large numbers of cells, perhaps repeated cell transplantation on multiple occasions. Using hESC-derived cells for this purpose will probably be quite expensive, since generating sufficient cell numbers will require expensive cell culture conditions, growth factors, cytokines and complex manipulations.

On the other hand, if transplanted cells were needed in only small numbers, as for reseeding of the liver, therapeutic applications using stem cells could be more readily contemplated. Examples of this situation will concern replacement of deficient proteins in monogenetic disorders without organ injury, e.g. disorders of bilirubin metabolism, genetic hypercholesterolaemia, glycogen storage diseases, hyperammonaemia syndromes, bile transport disorders, etc., as well as disorders leading to significant liver damage, e.g. Wilson disease with copper toxicosis[1]. Initial cell transplantation studies using mature hepatocytes in many such conditions have been encouraging, and offer an appropriate framework for further work[64].

Reconstitution of non-hepatic cell compartments, e.g. liver sinusoidal endothelium and hepatic stellate cells, constitutes further areas of interest. For instance, recent studies have demonstrated that hepatic stellate cells are responsible for producing the protease, ADAMTS13, which is deficient in thrombotic thrombocytopenic purpura (TTP)[65]. This condition is characterized by the development of arteriolar and capillary thrombi containing von Willebrand factor (vWF) and platelets. As vWF mediates platelet adhesion and aggregation at sites of vascular injury, abnormally high circulating levels of this glycoprotein are deleterious. This arises from the deficiency of ADAMTS13, a metalloproteinase that normally cleaves vWF. Whether replacement of the hepatic stellate cell compartment with either mature cells or stem cells will offer therapeutic opportunities in TTP needs to be considered. Similarly, as hepatic stellate cells are thought to play significant pathophysiological roles in liver fibrosis, perturbation of this compartment by cell and/or gene therapy approaches might be beneficial in altering the natural history of chronic liver diseases. Recent studies have demonstrated that hepatic stellate cells can be successfully targeted to hepatic sinusoidal endothelium and transplanted cells can survive in the liver over the long term[66]. Moreover, the possibility of hepatic stem cells themselves possessing stem cell potential has been advanced in recent studies[67].

Another exciting development concerns identification of liver sinusoidal endothelial cells as a major site for factor VIII (FVIII) production, which is deficient in haemophilia A[68]. At present, treating people with haemophilia A requires repeated administration of pooled plasma or more expensive recombinant FVIII. Previous studies in people with hemophilia A undergoing OLT showed prompt restoration of plasma FVIII levels after liver replacement[69]. However, the specific cell type that produced FVIII remained elusive because suitable materials were lacking to verify the presence of FVIII at the protein level. The candidate populations included hepatocytes, where FVIII mRNA was expressed at high levels, as well as liver sinusoidal endothelial cells and Kupffer cells, which also expressed FVIII, although at lower levels. However, use of short-term transplantation assays, in which healthy mouse cells were transplanted in the peritoneal cavity of haemophilia A knockout mice, have established that the relevant cell type capable of reconstituting FVIII deficiency was the liver sinusoidal endothelial cell and not the hepatocyte[68]. Further studies have demonstrated that replacement of diseased sinusoidal endothelial cells in the liver itself, by transplantation of healthy cells, cured haemophilia A in mice[70]; therefore hepatic sinusoidal endothelium represents another suitable target for stem cell-based therapies.

CONCLUSIONS

Emerging insights into the biology of multiple types of stem/progenitor cells, including embryonic, fetal and adult cells, offer unique opportunities for cell and gene therapy. Successful replacement of native cells with transplanted cells has the potential to generate novel animal models of disease. Moreover, reconstitution of specific hepatic cell compartments will be helpful in therapies aimed at various disorders, in which hepatocytes, liver sinusoidal endothelial cells, hepatic stellate cells, or other liver cells could be suitably targeted. The overall combination of insights into stem cell biology, liver development and homeostasis, and disease pathophysiology will help in obtaining rational strategies for applications of stem/progenitor cell transplantation. The most immediate applications of such efforts will probably emerge from the use of fetal stem/progenitor cells, which are already advanced along hepatobiliary lineages, rather than hESC-derived cells. On the other hand, significant uncertainties have emerged with respect to the hepatic potential of extrahepatic stem/progenitor cells, including bone marrow-derived haematopoietic stem cells and MAPC, although additional types of cells – including stem/progenitor cells from the amniotic fluid, placenta, or other sources – need further examination.

Acknowledgements

This work was supported in part by NIH grants R01 DK-46952, R01-DK-071111, P01-DK-052956, P30 DK-41296, and M01 RR-12248.

References

1. Gupta S. The therapeutic potential of liver repopulation for metabolic or endocrine disorders. In: Lester LB, editor. Stem Cells in Endocrinology. New Jersey: Humana Press, 2005:165–81.
2. Grossman M, Rader DJ, Muller DWM et al. A pilot study of *ex vivo* gene therapy for homozygous familiar hypercholesterolaemia. Nat Med. 1995;1:1148–54.
3. Wu YM, Joseph B, Gupta S. Immunosuppression using the mTOR inhibition mechanism affects replacement of the rat liver with transplanted cells. Hepatology. 2006;44:410–19.
4. Nagata H, Nishitai R, Shirota C et al. Prolonged survival of porcine hepatocytes in cynomolgus monkeys. Gastroenterology. 2007;132:321–9.
5. Gupta S. Hog heaven on the road to liver cell therapy. Gastroenterology. 2007;32:450–3.
6. Thomson JA, Itskovitz-Eldor J, Shapiro SS et al. Embryonic stem cell lines derived from human blastocysts. Science. 1998;282:1145–7.
7. Klimanskaya I, Chung Y, Meisner L, Johnson J, West MD, Lanza R. Human embryonic stem cells derived without feeder cells. Lancet. 2005;365:1636–41.
8. Shirahashi H, Wu J, Yamamoto N et al. Differentiation of human and mouse embryonic stem cells along a hepatocyte lineage. Cell Transplant. 2004;13:197–211.
9. Soto-Gutierrez A, Kobayashi N, Rivas-Carrillo JD et al. Reversal of mouse hepatic failure using an implanted liver-assist device containing ES cell-derived hepatocytes. Nat Biotechnol. 2006;24:1412–19.
10. Fausto N. Liver regeneration and repair: hepatocytes, progenitor cells, and stem cells. Hepatology. 2004;39:1477–87.
11. Alison MR. Characterization of the differentiation capacity of rat-derived hepatic stem cells. Semin Liver Dis. 2003;23:325–36.
12. Zhao R, Duncan SA. Embryonic development of the liver. Hepatology. 2005;41:956–67.
13. Overturf K, al-Dhalimy M, Ou CN, Finegold M, Grompe M. Serial transplantation reveals the stem-cell-like regenerative potential of adult mouse hepatocytes. Am J Pathol. 1997; 151:1273–80.
14. Badve S, Logdberg L, Sokhi R et al. An antigen reacting with das-1 monoclonal antibody is ontogenically regulated in diverse organs including liver and indicates sharing of developmental mechanisms among cell lineages. Pathobiology. 2000;68:76–86.
15. Malhi H, Irani AN, Gagandeep S, Gupta S. Isolation of human progenitor liver epithelial cells with extensive replication capacity and differentiation into mature hepatocytes. J Cell Sci. 2002;115:2679–88.
16. Lazaro CA, Croager EJ, Mitchell C et al. Establishment, characterization, and long-term maintenance of cultures of human fetal hepatocytes. Hepatology. 2003;38:1095–106.
17. Dan YY, Riehle KJ, Lazaro C et al. Isolation of multipotent progenitor cells from human fetal liver capable of differentiating into liver and mesenchymal lineages. Proc Natl Acad Sci USA. 2006;103:9912–17.
18. Dabeva MD, Petkov PM, Sandhu J et al. Proliferation and differentiation of fetal liver epithelial progenitor cells after transplantation into adult rat liver. Am J Pathol. 2000;156: 2017–31.
19. Sierra E, Maganto P, Codesal J et al. Liver gene expression and increase in albumin synthesis by fetal hepatocytes transplanted into analbuminemic rats. Life Sci. 2000;67: 2417–32.

20. Sandhu, JS, Petkov PM, Dabeva MD, Shafritz DA. Stem cell properties and repopulation of the rat liver by fetal liver epithelial progenitor cells. Am J Pathol. 2001;159:1323–34.
21. Cantz T, Zuckerman DM, Burda MR et al. Quantitative gene expression analysis reveals transition of fetal liver progenitor cells to mature hepatocytes after transplantation in uPA/RAG-2 mice. Am J Pathol. 2003;162:37–45.
22. Nierhoff D, Ogawa A, Oertel M, Chen YQ, Shafritz DA. Purification and characterization of mouse fetal liver epithelial cells with high *in vivo* repopulation capacity. Hepatology. 2005;42:130–9
23. Habibullah CM. Hepatocyte transplantation: need for liver cell bank. Trop Gastroenterol. 1992;13:129–31.
24. Khan AA, Habeeb A, Parveen N et al. Peritoneal transplantation of human fetal hepatocytes for the treatment of acute fatty liver of pregnancy: a case report. Trop Gastroenterol. 2004;25:141–3.
25. Shamblott MJ, Axelman J, Littlefield JW et al. Human embryonic germ cell derivatives express a broad range of developmentally distinct markers and proliferate extensively *in vitro*. Proc Natl Acad Sci USA. 2001;98:113–18.
26. Chinzei R, Tanaka Y, Shimizu-Saito K et al. Embryoid-body cells derived from a mouse embryonic stem cell line show differentiation into functional hepatocytes. Hepatology. 2002;36:22–9.
27. Yin Y, Lim YK, Salto-Tellez M, Ng SC, Lin CS, Lim SK. AFP(+), ESC-derived cells engraft and differentiate into hepatocytes *in vivo*. Stem Cells. 2002;20:338–46.
28. Ishizaka S, Shiroi A, Kanda S et al. Development of hepatocytes from ES cells after transfection with the HNF-3beta gene. FASEB J. 2002;16:1444–6.
29. Rambhatla L, Chiu CP, Kundu P, Peng Y, Carpenter MK. Generation of hepatocyte-like cells from human embryonic stem cells. Cell Transplant. 2003;12:1–11.
30. Lavon N, Yanuka O, Benvenisty N. Differentiation and isolation of hepatic-like cells from human embryonic stem cells. Differentiation. 2004;72:230–8.
31. Teratani T, Yamamoto H, Aoyagi K et al. Direct hepatic fate specification from mouse embryonic stem cells. Hepatology. 2005; 41:836–46.
32. Kubota H, Reid LM. Clonogenic hepatoblasts, common precursors for hepatocytic and biliary lineages, are lacking classical major histocompatibility complex class I antigen. Proc Natl Acad Sci USA. 2000;97:12132–7.
33. Suzuki A, Zheng YW, Fukao K, Nakauchi H, Taniguchi H. Liver repopulation by c-Met-positive stem/progenitor cells isolated from the developing rat liver. Hepatogastroenterology. 2004;51:423–6.
34. Zahler MH, Irani A, Malhi H et al. The application of a lentiviral vector for gene transfer in fetal human hepatocytes. J Gene Med. 2000;2:186–93.
35. Wege H, Le HT, Chui MS et al. Telomerase reconstitution immortalizes human fetal hepatocytes without disrupting their differentiation potential. Gastroenterology. 2003;124: 432–44.
36. Zalzman M, Gupta S, Giri RK et al. Reversal of hyperglycemia in mice using human expandable insulin-producing cells differentiated from fetal liver progenitor cells. Proc Natl Acad Sci USA. 2003;100:7253–8.
37. Zalzman M, Anker-Kitai L, Efrat S. Differentiation of human liver-derived, insulin-producing cells toward the beta-cell phenotype. Diabetes. 2005;54:2568–75.
38. Qui C, Hanson E, Olivier E et al. Differentiation of human embryonic stem cells into hematopoietic cells by coculture with human fetal liver cells recapitulates the globin switch that occurs early in development. Exp Hematol. 2005;33:1450–8.
39. Saxena R, Theise N. Canals of Hering: recent insights and current knowledge. Semin Liver Dis. 2004;24:43–8.
40. Yasui O, Miura N, Terada K, Kawarada Y, Koyama K, Sugiyama T. Isolation of oval cells from Long-Evans Cinnamon rats and their transformation into hepatocytes *in vivo* in the rat liver. Hepatology. 1997;25:329–34.
41. Petersen BE, Bowen WC, Patrene KD et al. Bone marrow as a potential source of hepatic oval cells. Science. 1999;284:1168–70.
42. Theise ND, Badve S, Saxena R et al. Derivation of hepatocytes from bone marrow cells in mice after radiation-induced myeloablation. Hepatology. 2000;31:235–40.
43. Alison MR, Poulsom R, Jeffery R et al. Hepatocytes from non-hepatic adult stem cells. Nature. 2000;406:257.

44. Lagasse E, Connors H, Al Dhalimy M et al. Purified hematopoietic stem cells can differentiate into hepatocytes *in vivo*. Nat Med. 2000;6:1229–34.
45. Jiang Y, Jahagirdar BN, Reinhardt RL et al. Pluripotency of mesenchymal stem cells derived from adult marrow. Nature. 2002;418:41–9.
46. Kakinuma S, Tanaka Y, Chinzei R et al. Human umbilical cord blood as a source of transplantable hepatic progenitor cells. Stem Cells. 2003;21:217–27.
47. Wang X, Ge S, McNamara G, Hao QL, Crooks GM, Nolta JA. Albumin-expressing hepatocyte-like cells develop in the livers of immune-deficient mice that received transplants of highly purified human hematopoietic stem cells. Blood. 2003;101:4201–8
48. Newsome PN, Johannessen I, Boyle S et al. Human cord blood-derived cells can differentiate into hepatocytes in the mouse liver with no evidence of cellular fusion. Gastroenterology. 2003;124:1891–900.
49. Tanabe Y, Tajima F, Nakamura Y et al. Analyses to clarify rich fractions in hepatic progenitor cells from human umbilical cord blood and cell fusion. Biochem Biophys Res Commun. 2004;324:711–18.
50. Turrini P, Monego G, Gonzalez J et al. Human hepatocytes in mice receiving pre-immune injection with human cord blood cells. Biochem Biophys Res Commun. 2005;326:66–73.
51. Terada S, Matsuura K, Enosawa S et al. Inducing proliferation of human amniotic epithelial (HAE) cells for cell therapy. Cell Transplant. 2000;9:701–4.
52. Takashima S, Ise H, Zhao P, Akaike T, Nikaido T. Human amniotic epithelial cells possess hepatocyte-like characteristics and functions. Cell Struct Funct. 2004;29:73–84.
53. Yen BL, Huang HI, Chien CC et al. Isolation of multipotent cells from human term placenta. Stem Cells. 2005;23:3–9.
54. Wang X, Montini E, Al-Dhalimy M, Lagasse E, Finegold M, Grompe M. Kinetics of liver repopulation after bone marrow transplantation. 2002;161:565–74.
55. Krause DS, Theise ND, Collector MI et al. Multi-organ, multi-lineage engraftment by a single bone marrow-derived stem cell. Cell. 2001;105:369–77.
56. Keene CD, Ortiz-Gonzalez XR, Jiang Y, Largaespada DA, Verfaillie CM, Low WC. Neural differentiation and incorporation of bone marrow-derived multipotent adult progenitor cells after single cell transplantation into blastocyst stage mouse embryos. Cell Transplant. 2003;12:201–13.
57. Wagers AJ, Sherwood RI, Christensen JL, Weissman IL. Little evidence for developmental plasticity of adult hematopoietic stem cells. Science. 2002;297:2256–9.
58. Kanazawa Y, Verma IM. Little evidence of bone marrow-derived hepatocytes in the replacement of injured liver. Proc Natl Acad Sci USA. 2003;100(Suppl. 1):11850–3.
59. Cantz T, Sharma AD, Jochheim-Richter A et al. Reevaluation of bone marrow-derived cells as a source for hepatocyte regeneration. Cell Transplant. 2004;13:659–66.
60. Vassilopoulos G, Wang PR, Russell DW. Transplanted bone marrow regenerates liver by cell fusion. Nature. 2003;422:901–4.
61. Willenbring H, Bailey AS, Foster M et al. Myelomonocytic cells are sufficient for therapeutic cell fusion in liver. Nat Med. 2004;10:744–8.
62. Jang YY, Collector MI, Baylin SB, Diehl AM, Sharkis SJ. Hematopoietic stem cells convert into liver cells within days without fusion. Nat Cell Biol. 2004;6:532–9.
63. Kajstura J, Rota M, Whang B et al. Bone marrow cells differentiate in cardiac cell lineages after infarction independently of cell fusion. Circ Res. 2005;96:127–37.
64. Fisher RA, Strom SC. Human hepatocyte transplantation: worldwide results. Transplantation. 2006;82:441–9.
65. Zhou W, Inada M, Lee TP et al. ADAMTS13 is expressed in hepatic stellate cells. Lab Invest. 2005;85:780–8.
66. Benten D, Kumaran V, Bhargava KK, Palestro CJ, Gupta S. Radiolabeling of hepatic stellate cells with In-111-oxine enables biodistribution analysis for cell transplantation. Eur J Nucl Med. 2004;31:424 (Abstract).
67. Sicklick JK, Li YX, Melhem A et al. Hedgehog signaling maintains resident hepatic progenitors throughout life. Am J Physiol Gastrointest Liver Physiol. 2006; 291:G575–83.
68. Kumaran V, Benten D, Follenzi A, Joseph B, Sarkar R, Gupta S. Transplantation of endothelial cells corrects the phenotype in hemophilia A mice. J Thromb Haemost. 2005;3: 2022–31.
69. Bontempo FA, Lewis JH, Gorenc TJ et al. Liver transplantation in hemophilia A. Blood. 1987;69:1721–4.

70. Follenzi A, Ulla M, Novikoff P, Gupta S. Reconstitution of the mouse liver with genetically modified liver sinusoidal endothelial cells (LSEC) for biological studies and cell and gene therapy applications. Hepatology. 2006;44:255A (Abstract).

Index

Falk Symposium Series

132. Galle PR, Gerken G, Schmidt WE, Wiedenmann B, eds. *Disease Progression and Carcinogenesis in the Gastrointestinal Tract*. Falk Symposium 132. 2003 ISBN 0-7923-8785-6
132A. Staritz M, Adler G, Knuth A, Schmiegel W, Schmoll H-J, eds. *Side-effects of Chemotherapy on the Gastrointestinal Tract*. Falk Workshop. 2003 ISBN 0-7923-8791-0
132B. Reutter W, Schuppan D, Tauber R, Zeitz M, eds. *Cell Adhesion Molecules in Health and Disease.* Falk Workshop. 2003 ISBN 0-7923-8786-4
133. Duchmann R, Blumberg R, Neurath M, Schölmerich J, Strober W, Zeitz M. *Mechanisms of Intestinal Inflammation: Implications for Therapeutic Intervention in IBD.* Falk Symposium 133. 2004 ISBN 0-7923-8787-2
134. Dignass A, Lochs H, Stange E. *Trends and Controversies in IBD – Evidence-Based Approach or Individual Management?* Falk Symposium 134. 2004 ISBN 0-7923-8788-0
134A. Dignass A, Gross HJ, Buhr V, James OFW. *Topical Steroids in Gastroenterology and Hepatology.* Falk Workshop. 2004 ISBN 0-7923-8789-9
135. Lukáš M, Manns MP, Špičák J, Stange EF, eds. *Immunological Diseases of Liver and Gut.* Falk Symposium 135. 2004 ISBN 0-7923-8792-9
136. Leuschner U, Broomé U, Stiehl A, eds. *Cholestatic Liver Diseases: Therapeutic Options and Perspectives.* Falk Symposium 136. 2004 ISBN 0-7923-8793-7
137. Blum HE, Maier KP, Rodés J, Sauerbruch T, eds. *Liver Diseases: Advances in Treatment and Prevention.* Falk Symposium 137. 2004 ISBN 0-7923-8794-5
138. Blum HE, Manns MP, eds. *State of the Art of Hepatology: Molecular and Cell Biology.* Falk Symposium 138. 2004 ISBN 0-7923-8795-3
138A. Hayashi N, Manns MP, eds. *Prevention of Progression in Chronic Liver Disease: An Update on SNMC (Stronger Neo-Minophagen C).* Falk Workshop. 2004 ISBN 0-7923-8796-1
139. Adler G, Blum HE, Fuchs M, Stange EF, eds. *Gallstones: Pathogenesis and Treatment.* Falk Symposium 139. 2004 ISBN 0-7923-8798-8
140. Colombel J-F, Gasché C, Schölmerich J, Vucelic C, eds. *Inflammatory Bowel Disease: Translation from Basic Research to Clinical Practice.* Falk Symposium 140. 2005. ISBN 1-4020-2847-4
141. Paumgartner G, Keppler D, Leuschner U, Stiehl A, eds. *Bile Acid Biology and its Therapeutic Implications.* Falk Symposium 141. 2005 ISBN 1-4020-2893-8
142. Dienes H-P, Leuschner U, Lohse AW, Manns MP, eds. *Autoimmune Liver Disease.* Falk Symposium 142. 2005 ISBN 1-4020-2894-6
143. Ammann RW, Büchler MW, Adler G, DiMagno EP, Sarner M, eds. *Pancreatitis: Advances in Pathobiology, Diagnosis and Treatment.* Falk Symposium 143. 2005 ISBN 1-4020-2895-4
144. Adler G, Blum AL, Blum HE, Leuschner U, Manns MP, Mössner J, Sartor RB, Schölmerich J, eds. *Gastroenterology Yesterday – Today – Tomorrow: A Review and Preview.* Falk Symposium 144. 2005 ISBN 1-4020-2896-2
145. Henne-Bruns D, Buttenschön K, Fuchs M, Lohse AW, eds. *Artificial Liver Support.* Falk Symposium 145. 2005 ISBN 1-4020-3239-0
146. Blumberg RS, Gangl A, Manns MP, Tilg H, Zeitz M, eds. *Gut–Liver Interactions: Basic and Clinical Concepts.* Falk Symposium 146. 2005 ISBN 1-4020-4143-8
147. Jewell DP, Colombel JF, Peña AS, Tromm A, Warren BS, eds. *Colitis: Diagnosis and Therapeutic Strategies.* Falk Symposium 147. 2006 ISBN 1-4020-4315-5

Falk Symposium Series

148. Kruis W, Forbes A, Jauch K-W, Kreis ME, Wexner SD, eds. *Diverticular Disease: Emerging Evidence in a Common Condition.* Falk Symposium 148. 2006 ISBN 1-4020- 4317-1
149. van Cutsem E, Rustgi AK, Schmiegel W, Zeitz M, eds. *Highlights in Gastrointestinal Oncology.* Falk Symposium 149. 2006. ISBN 1-4020-5108-5
150. Galle PR, Gerken G, Schmidt WE, Wiedenmann B, eds. *Disease Progression and Disease Prevention in Hepatology and Gastroenterology.* Falk Symposium 150. 2006 ISBN 1-4020-5109-3
151. Fraser A, Gibson PR, Hibi T, Qian J-M, Schölmerich, eds. *Emerging Issues in Inflammatory Bowel Disease.* Falk Symposium 151. 2006 ISBN-13 978-1-4020-5701-4
152. Not published.
153. Dignass A, Rachmilewitz D, Stange E-F, Weinstock JV, eds. *Immunoregulation in Inflammatory Bowel Diseases – Current Understanding and Innovation.* Falk Symposium 153. 2007 ISBN-13 978-1-4020-5888-2
154. Adler G, Fiocchi C, Lazebnik LB, Vorobiev GI, eds. *Inflammatory Bowel Disease – Diagnostic and Therapeutic Strategies.* Falk Symposium 154. 2007 ISBN-13 978-1-4020-6115-8
155. Keppler D, Beuers U, Leuschner U, Stiehl A, Trauner M, Paumgartner G, eds. *Bile Acids: Biological Actions and Clinical Relevance.* Falk Symposium 155. 2007 ISBN 978-1-4020-6251-3
156. Blum HE, Cox DW, Häussinger D, Jansen PLM, Kullak-Ublick GA, eds. *Genetics in Liver Diseases.* Falk Symposium 156. 2007 ISBN 978-14020-6393-0